APPLICATIONS OF CALCULUS

Philip Straffin, Editor

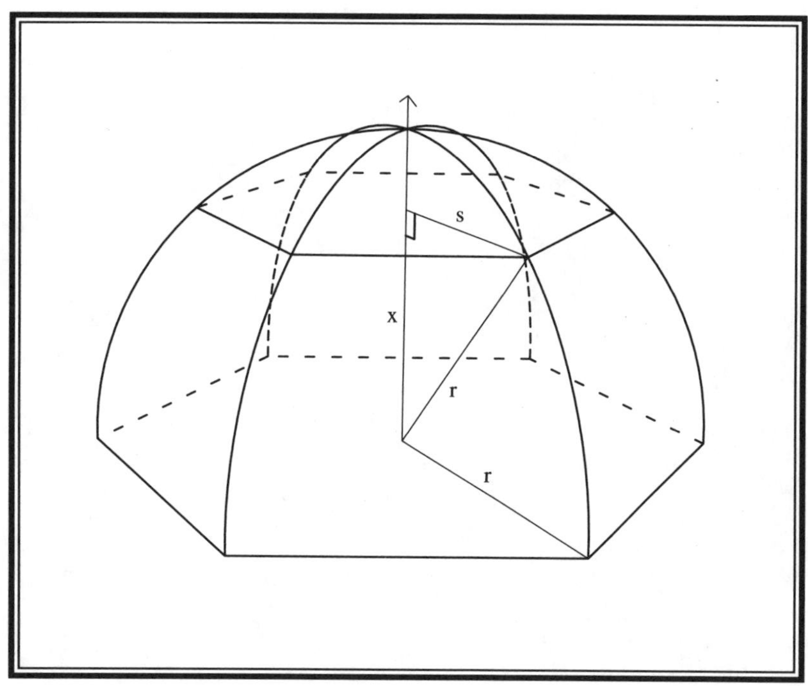

Resources for Calculus Collection

Volume 3

APPLICATIONS OF CALCULUS

Philip Straffin, Editor

A Project of
The Associated Colleges of the Midwest and
The Great Lakes Colleges Association

Writers for this Volume

Clark Benson
National Security Agency

Paul Campbell
Beloit College

Kevin Hastings
Knox College

Steven Janke
Colorado College

Walter Meyer
Adelphi University

Thomas Moore
Grinnell College

Philip Straffin
Beloit College

Supported by the National Science Foundation
A. Wayne Roberts, Project Director

MAA Notes Volume 29

Published and Distributed by
The Mathematical Association of America

MAA Notes and Reports Series

The MAA Notes and Reports Series, started in 1982, addresses a broad range of topics and themes of interest to all who are involved with undergraduate mathematics. The volumes in this series are readable, informative, and useful, and help the mathematical community keep up with developments of importance to mathematics.

MAA Notes

1. Problem Solving in the Mathematics Curriculum, *Committee on the Teaching of Undergraduate Mathematics,* a subcommittee of the Committee on the Undergraduate Program in Mathematics, *Alan H. Schoenfeld,* Editor
2. Recommendations on the Mathematical Preparation of Teachers, *Committee on the Undergraduate Program in Mathematics, Panel on Teacher Training.*
3. Undergraduate Mathematics Education in the People's Republic of China, *Lynn A. Steen,* Editor.
5. American Perspectives on the Fifth International Congress on Mathematical Education, *Warren Page,* Editor.
6. Toward a Lean and Lively Calculus, *Ronald G. Douglas,* Editor.
8. Calculus for a New Century, *Lynn A. Steen,* Editor.
9. Computers and Mathematics: The Use of Computers in Undergraduate Instruction, *Committee on Computers in Mathematics Education, D. A. Smith, G. J. Porter, L. C. Leinbach, and R. H. Wenger,* Editors.
10. Guidelines for the Continuing Mathematical Education of Teachers, *Committee on the Mathematical Education of Teachers.*
11. Keys to Improved Instruction by Teaching Assistants and Part-Time Instructors, *Committee on Teaching Assistants and Part-Time Instructors, Bettye Anne Case,* Editor.
13. Reshaping College Mathematics, *Committee on the Undergraduate Program in Mathematics, Lynn A. Steen,* Editor.
14. Mathematical Writing, by *Donald E. Knuth, Tracy Larrabee, and Paul M. Roberts.*
15. Discrete Mathematics in the First Two Years, *Anthony Ralston,* Editor.
16. Using Writing to Teach Mathematics, *Andrew Sterrett,* Editor.
17. Priming the Calculus Pump: Innovations and Resources, *Committee on Calculus Reform and the First Two Years,* a subcomittee of the Committee on the Undergraduate Program in Mathematics, *Thomas W. Tucker,* Editor.
18. Models for Undergraduate Research in Mathematics, *Lester Senechal,* Editor.
19. Visualization in Teaching and Learning Mathematics, *Committee on Computers in Mathematics Education, Steve Cunningham and Walter S. Zimmermann,* Editors.
20. The Laboratory Approach to Teaching Calculus, *L. Carl Leinbach et al.,* Editors.
21. Perspectives on Contemporary Statistics, *David C. Hoaglin and David S. Moore,* Editors.
22. Heeding the Call for Change: Suggestions for Curricular Action, *Lynn A. Steen,* Editor.
23. Statistical Abstract of Undergraduate Programs in the Mathematical Sciences and Computer Science in the United States: 1990–91 CBMS Survey, *Donald J. Albers, Don O. Loftsgaarden, Donald C. Rung, and Ann E. Watkins.*
24. Symbolic Computation in Undergraduate Mathematics Education, *Zaven A. Karian,* Editor.
25. The Concept of Function: Aspects of Epistemology and Pedagogy, *Guershon Harel and Ed Dubinsky,* Editors.

26. Statistics for the Twenty-First Century, *Florence and Sheldon Gordon,* Editors.
27. Resources for Calculus Collection, Volume 1: Learning by Discovery: A Lab Manual for Calculus, *Anita E. Solow,* Editor.
28. Resources for Calculus Collection, Volume 2: Calculus Problems for a New Century, *Robert Fraga,* Editor.
29. Resources for Calculus Collection, Volume 3: Applications of Calculus, *Philip Straffin,* Editor.
30. Resources for Calculus Collection, Volume 4: Problems for Student Investigation, *Michael B. Jackson and John R. Ramsay,* Editors.
31. Resources for Calculus Collection, Volume 5: Readings for Calculus, *Underwood Dudley,* Editor.

MAA Reports

1. A Curriculum in Flux: Mathematics at Two-Year Colleges, *Subcommittee on Mathematics Curriculum at Two-Year Colleges,* a joint committee of the MAA and the American Mathematical Association of Two-Year Colleges, *Ronald M. Davis,* Editor.
2. A Source Book for College Mathematics Teaching, *Committee on the Teaching of Undergraduate Mathematics, Alan H. Schoenfeld,* Editor.
3. A Call for Change: Recommendations for the Mathematical Preparation of Teachers of Mathematics, *Committee on the Mathematical Education of Teachers, James R. C. Leitzel,* Editor.
4. Library Recommendations for Undergraduate Mathematics, *CUPM ad hoc Subcommittee, Lynn A. Steen,* Editor.
5. Two-Year College Mathematics Library Recommendations, *CUPM ad hoc Subcommittee, Lynn A. Steen,* Editor.

These volumes may be ordered from the Mathematical Association of America,
1529 Eighteenth Street, NW, Washington, DC 20036.
202-387-5200 FAX 202-265-2384

Third Printing

ISBN 0-88385-085-0
Library of Congress Catalog Number 92-62281
Printed in the United States of America
Current Printing
10 9 8 7 6 5 4 3

INTRODUCTION
RESOURCES FOR CALCULUS COLLECTION

Beginning with a conference at Tulane University in January, 1986, there developed in the mathematics community a sense that calculus was not being taught in a way befitting a subject that was at once the culmination of the secondary mathematics curriculum and the gateway to collegiate science and mathematics. Far too many of the students who started the course were failing to complete it with a grade of C or better, and perhaps worse, an embarrassing number who did complete it professed either not to understand it or not to like it, or both. For most students it was not a satisfying culmination of their secondary preparation, and it was not a gateway to future work. It was an exit.

Much of the difficulty had to do with the delivery system: classes that were too large, senior faculty who had largely deserted the course, and teaching assistants whose time and interest were focused on their own graduate work. Other difficulties came from well intentioned efforts to pack into the course all the topics demanded by the increasing number of disciplines requiring calculus of their students. It was acknowledged, however, that if the course had indeed become a blur for students, it just might be because those choosing the topics to be presented and the methods for presenting them had not kept their goals in focus.

It was to these latter concerns that we responded in designing our project. We agreed that there ought to be an opportunity for students to discover instead of always being told. We agreed that the availability of calculators and computers not only called for exercises that would not be rendered trivial by such technology, but would in fact direct attention more to ideas than to techniques. It seemed to us that there should be explanations of applications of calculus that were self-contained, and both accessible and relevant to students. We were persuaded that calculus students should, like students in any other college course, have some assignments that called for library work, some pondering, some imagination, and above all, a clearly reasoned and written conclusion. Finally, we came to believe that there should be available to students some collateral readings that would set calculus in an intellectual context.

We reasoned that the achievement of these goals called for the availability of new materials, and that the uncertainty of just what might work, coupled with the number of people trying to address the difficulties, called for a large collection of materials from which individuals could select. Our goal was to develop such materials, and to encourage people to use them in any way they saw fit. In this spirit, and with the help of the Notes editor and committee of the Mathematical Association of America, we have produced five volumes of materials that are, with the exception of volume V where we do not hold original copyrights, meant to be in the public domain.

We expect that some of these materials may be copied directly and handed to an entire class, while others may be given to a single student or group of students. Some will provide a basis from which local adaptations can be developed. We will be pleased if authors ask for permission, which we expect to be generous in granting, to incorporate our materials into texts or laboratory manuals. We hope that in all of these ways, indeed in any way short of reproducing substantial segments to sell for profit, our material will be used to greatly expand ideas about how the calculus might be taught.

Though I as Project Director never entertained the idea that we could write a single text that would be acceptable to all 26 schools in the project, it was clear that some common notion of topics essential to any calculus course would be necessary to give us direction. The task of forging a common syllabus was managed by Andy Sterrett with a tact and efficiency that was instructive to us all, and the product of this work, an annotated core syllabus, appears as an appendix in Volume 1. Some of the other volumes refer to this syllabus to indicate where, in a course, certain materials might be used.

This project was situated in two consortia of liberal arts colleges, not because we intended to develop materials for this specific audience, but because our schools provide a large reservoir of classroom teachers who lavish on calculus the same attention a graduate faculty might give to its introductory analysis course. Our schools, in their totality, were equipped with most varieties of computer labs, and we included in our consortia many people who had become national leaders in the use of computer algebra systems.

We also felt that our campuses gave us the capability to test materials in the classroom. The size of our schools enables us to implement a new idea without cutting through the red tape of a larger institution, and we can just as quickly reverse ourselves when it is apparent that what we are doing is not working. We are practiced in going in both directions. Continual testing of the materials we were developing was seen as an integral part of our project, an activity that George Andrews, with the title of Project Evaluator, kept before us throughout the project.

The value of our contributions will now be judged by the larger mathematical community, but I was right in thinking that I could find in our consortia the great abundance of talent necessary for an undertaking of this magnitude. Anita Solow brought to the project a background of editorial work and quickly became not only one of the editors of our publications, but also a person to whom I turned for advice regarding the project as a whole. Phil Straffin, drawing on his association with UMAP, was an ideal person to edit a collection of applications, and was another person who brought editorial experience to our project. Woody Dudley came to the project as a writer well known for his witty and incisive commentary on mathematical literature, and was an ideal choice to assemble a collection of readings.

Our two editors least experienced in mathematical exposition, Bob Fraga and Mic Jackson, both justified the confidence we placed in them. They brought to the project an enthusiasm and freshness from which we all benefited, and they were able at all points in the project to draw upon an excellent corps of gifted and experienced writers. When, in the last months of the project, Mic Jackson took an overseas assignment on an Earlham program, it was possible to move John Ramsay into Mic's position precisely because of the excellent working relationship that had existed on these writing teams.

The entire team of five editors, project evaluator and syllabus coordinator worked together as a harmonious team over the five year duration of this project. Each member, in turn, developed a group of writers, readers, and classroom users as necessary to complete the task. I believe my chief contribution was to identify and bring these talented people together, and to see that they were supported both financially and by the human resources available in the schools that make up two remarkable consortia.

A. Wayne Roberts
Macalester College
1993

THE FIVE VOLUMES OF THE RESOURCES FOR CALCULUS COLLECTION

1. Learning by Discovery: A Lab Manual for Calculus
Anita E. Solow, editor

The availability of electronic aids for calculating makes it possible for students, led by good questions and suggested experiments, to discover for themselves numerous ideas once accessible only on the basis of theoretical considerations. This collection provides questions and suggestions on 26 different topics. Developed to be independent of any particular hardware or software, these materials can be the basis of formal computer labs or homework assignments. Although designed to be done with the help of a computer algebra system, most of the labs can be successfully done with a graphing calculator.

2. Calculus Problems for a New Century
Robert Fraga, editor

Students still need drill problems to help them master ideas and to give them a sense of progress in their studies. A calculator can be used in many cases, however, to render trivial a list of traditional exercises. This collection, organized by topics commonly grouped in sections of a traditional text, seeks to provide exercises that will accomplish the purposes mentioned above, even for the student making intelligent use of technology.

3. Applications of Calculus
Philip Straffin, editor

Everyone agrees that there should be available some self-contained examples of applications of the calculus that are tractable, relevant, and interesting to students. Here they are, 18 in number, in a form to be consulted by a teacher wanting to enrich a course, to be handed out to a class if it is deemed appropriate to take a day or two of class time for a good application, or to be handed to an individual student with interests not being covered in class.

4. Problems for Student Investigation
Michael B. Jackson and John R. Ramsay, editors

Calculus students should be expected to work on problems that require imagination, outside reading and consultation, cooperation, and coherent writing. They should work on open-ended problems that admit several different approaches and call upon students to defend both their methodology and their conclusion. Here is a source of 30 such projects.

5. Readings for Calculus
Underwood Dudley, editor

Faculty members in most disciplines provide students in beginning courses with some history of their subject, some sense not only of what was done by whom, but also of how the discipline has contributed to intellectual history. These essays, appropriate for duplicating and handing out as collateral reading aim to provide such background, and also to develop an understanding of how mathematicians view their discipline.

ACKNOWLEDGEMENTS

Besides serving as editors of the collections with which their names are associated, Underwood Dudley, Bob Fraga, Mic Jackson, John Ramsay, Anita Solow, and Phil Straffin joined George Andrews (Project Evaluator), Andy Sterrett (Syllabus Coordinator) and Wayne Roberts (Project Director) to form a steering committee. The activities of this group, together with the writers' groups assembled by the editors, were supported by two grants from the National Science Foundation.

The NSF grants also funded two conferences at Lake Forest College that were essential to getting wide participation in the consortia colleges, and enabled member colleges to integrate our materials into their courses.

The projects benefited greatly from the counsel of an Advisory Committee that consisted of Morton Brown, Creighton Buck, Jean Callaway, John Rigden, Truman Schwartz, George Sell, and Lynn Steen.

Macalester College served as the grant institution and fiscal agent for this project on behalf of the schools of the Associated Colleges of the Midwest (ACM) and Great Lakes Colleges Association (GLCA) listed below.

ACM	GLCA
Beloit College	Albion College
Carleton College	Antioch College
Coe College	Denison University
Colorado College	DePauw University
Cornell College	Earlham College
Grinnell College	Hope College
Knox College	Kalamazoo College
Lake Forest College	Kenyon College
Lawrence University	Oberlin College
Macalester College	Ohio Wesleyan University
Monmouth College	Wabash College
Ripon College	College of Wooster
St. Olaf College	
University of Chicago	

I would also like to thank Stan Wagon of Macalester College for providing the cover image for each volume in the collection.

ACKNOWLEDGMENTS

TABLE OF CONTENTS

Calculus II

Calculus of Several Variables

PREFACE: USING THESE MODULES

Many students leave a traditional calculus course with very little sense of the powerful role calculus plays in the modern world. They don't know that calculus is the language which describes anything that changes, that much of the technology they use is based on ideas from calculus, that calculus is useful in social science as well as physical and biological science, or that optimization is a key to efficiency in business and industrial processes. If we merely tell students these things, they may not take our pronouncements very seriously. If we offer students only "toy" applications, we *invite* them not to take us seriously.

Yet calculus *is* useful, and giving students access to the variety of fascinating applications of calculus can enrich their learning. That is the goal of this volume.

The eighteen applications collected here are diverse, and I hope they will be interesting to students even beyond a student's particular field of interest. I think students might be interested in ways to arbitrate disputes even if they aren't economists, in how fast a raindrop falls even if they aren't physicists, or in the spread of AIDS even if they aren't doctors. Each module starts with a concrete problem—Is a Canadian voting scheme fair? What happens when you tune a radio? How could you choose a best portfolio of stocks?—and develops a solution to the problem based on the ideas of calculus. The discussions are fairly detailed, realistic, and pay careful attention to the process of mathematical modeling. Students can learn a lot from them.

The applications are listed in order of the calculus ideas they use, and where those ideas appear in a standard calculus sequence. On pages 3–5 you will find the ACM-GLCA calculus project curriculum committee's suggested syllabi for Calculus I and two versions of Calculus II, annotated with suggestions for where particular application modules might fit. Note that some modules are listed in two places. For instance, "Moving a Planar Robot Arm" is listed as a Calculus II module because it deals with parametric motion in the plane. However, it is written so that students could understand it as soon as they understand the idea of a derivative, and it has been used successfully in Calculus I classes. In the other direction, a number of instructors have found that using "Calculus I" modules in Calculus II helps students review basic ideas. More detailed information on the prerequisites needed for each module can be found at the beginning of that module.

I want to say something more about the matter of difficulty. Applications, like the ideas of calculus itself, are hard. You have to think hard to understand a problem, to model that problem as mathematics, to solve it, and to understand what you have and haven't solved. This does not mean that students shouldn't learn about applications—if they can't understand a use of calculus, it probably doesn't do them much good to "know" calculus. It *does* mean that we should give them help, and provide an environment which is as friendly as possible. In particular, I think the following are necessary:

• You should read any module before you assign it to students. Judge if the level of difficulty is appropriate.

• You should be available to help students. Give class time, or invite them in to talk with you. For real problems, consulting experts is an accepted practice.

• You should let students help each other. I recommend, in fact, that the modules be worked on in groups (see below). Real applications usually involve teamwork, and talking mathematics benefits everyone involved.

With an open and helpful environment, our students can do impressive things.

These modules can be, and have been, used in many different ways. Here are some:

• The instructor uses modules as bases for occasional "application day" lectures.

• Modules are given to eager students, or students interested in particular application areas, to read for interest.

• Students report on modules to a math club, or department colloquium, or a special "calculus honors seminar."

• Students are asked to read a module and do exercises from it. This works best if a student is offered a choice of several modules, has considerable flexibility about which problems to do, and has many chances to get help.

• Students write papers based on modules. One format is to ask the student to find or construct a problem which could be solved by a particular technique, and solve it, explaining all steps carefully. Again, choice and ability to get help are important.

• Students give group reports or panel presentations about modules.

• Students read modules and do exercises in groups. The optimal size for a group seems to be three. I recommend giving class time for the groups to meet, with you there to answer questions. I use the modules this way, breaking class twice during the term, for a week, to work in groups on applications. It is an appreciated change of pace.

I hope your students enjoy these applications, and come away intrigued and convinced of the usefulness of calculus.

I am grateful to the writers of the modules, to Creighton Buck and Tom Barr who diligently read all of the modules and offered suggestions, to my students who worked on versions of these modules and told me what they liked and didn't, and to all the faculty and students in the ACM-GLCA liberal arts colleges who class-tested the modules in 1990-1992.

—Philip Straffin, Beloit College, Beloit, WI 53511

CALCULUS I: THE DERIVATIVE AND THE INTEGRAL

1. Introduction

2. Functions and Graphs
   - definition, domain and range, linear and quadratic functions
   - trigonometric functions (sine, cosine and tangent)
   - exponential and logarithmic functions
   - composite functions
   - functions described by tables and graphs

3. The Derivative
   - average rates of change
   - instantaneous rates of change, developed intuitively
   - a study of limits, either intuitively or epsilon-delta
   - definition and properties of the derivative
   - derivatives of polynomials, sine and cosine
   - derivatives of exponential and logarithmic functions
   - derivative of sums, differences, products and quotients
   - the chain rule and inverse functions

4. Extreme Values
   - extreme values; approximate graphical or numerical solutions
   - existence theorem for a function continuous on a closed, bounded interval
   - critical point theorem: extreme values are attained only at critical points
   - monotonicity theorem: a function with positive derivative is increasing
   - concavity theorem: a function with positive second derivative is concave up
   - first and second derivative test for local extremes
   - the mean value theorem

   1, 2, 3, 4, 5, 13, 14

5. Antiderivatives and Differential Equations
   - antiderivative and their basic properties
   - introduction to differential equations; separation of variables; constants of integration and initial conditions

   6, 7

6. The Definite Integral
   - Riemann sums
   - limit of Riemann sums
   - integrability theorem; properties of definite integrals
   - the fundamental theorem of calculus
   - the derivative of integrals with variable upper bounds

   8, 9, 10

CALCULUS IIA: EXACT AND APPROXIMATE REPRESENTATION OF NUMBERS AND FUNCTIONS

1. Introduction

2. The Integral Revisited
   - the definite integral: exact values from the fundamental theorem of calculus
   - antiderivatives: finding them by substitution, including trigonometric substitutions; integration by parts
   - the definite integral: approximate values by Riemann sums and the trapezoidal rule, with some error analysis

   7, 8, 9, 10, 11, 12, 15

3. Sequences and Series of Numbers
   Sequence topics:
   - infinite sequences as functions
   - limits of sequences
   - recursively defined sequences **5**
   - improper integrals; l'Hopital's rule **16**
   - limits at infinity and the asymptotic behavior of functions

   Series topics:
   - infinite series
   - geometric series
   - the nth term test for divergence
   - equivalence of series, and the limit comparison test
   - p-series, with emphasis on the harmonic series

4. Sequences and Series of Functions
   - the mean value theorem revisited and its second degree analogue
   - Taylor polynomials with remainder theorem
   - graphical comparison of a function and its Taylor polynomials; the graph of the error function for a Taylor approximation
   - error estimation on intervals
   - Taylor series: the general expansion and examples (sine, cosine, exponential, logarithmic, the binomial theorem)
   - power series, with ratio test to give domains of convergence
   - algebraic manipulation and term-by-term integration and differentiation

5. Series Solutions of Differential Equations
   - defining functions with differential equations, for example $y''+ky=0$ and $y'=ky$
   - solving homogeneous linear second order equations with constant coefficients using power series

CALCULUS IIB: CALCULUS IN A THREE-DIMENSIONAL WORLD

1. Introduction

2. The Integral Revisited
   - the definite integral: exact values from the fundamental theorem of calculus
   - antiderivatives: finding them by substitution, including trigonometric substitutions; integration by parts
   - the definite integral: approximate values by Riemann sums and the trapezoidal rule, with some error analysis

   7, 8, 9, 10, 11, 12, 15

3. The Integral in R^2 and R^3
   - real-values functions of two and three variables; graphing; level curves
   - definitions of double and triple integrals
   - integrals over rectangles and boxes
   - evaluation of double integrals over regions with curves boundaries

4. The Derivative in Two and Three Variables
   - partial derivatives: definition and geometric motivation
   - equation of the tangent plane
   - unconstrained optimization: critical points and the second derivative test
   - curves described by parametric equations **13, 14**
   - the chain rule
   - extreme value theorem revisited
   - constrained optimization; Lagrange multipliers **17, 18**

5. Integration Along Curves
   - definition of the Riemann integral of a real function on a curve in R^2 and R^3
   - vector fields in R^2 and R^3 and the dot product
   - line integrals
   - Green's theorem and path independence

ARBITRATING DISPUTES

Author: Philip Straffin, Beloit College, Beloit, WI 53511

Area of Application: economics

Calculus needed: derivative of polynomials, maximization on a closed interval.

Related mathematics: utility theory, axiomatic systems in social science.

An Arbitration Problem

Management and Labor are negotiating over a new contract. Each side has concessions it wishes to get from the other. Labor is asking for a one dollar per hour across-the-board raise and a package of increased pension benefits. Management is concerned that the fifteen minute morning coffee break is being abused—workers are straggling back late and the assembly line is being disrupted—and would like to eliminate it. Management would also like to automate one of the checkpoints on the line, which would eliminate eight union jobs.

Negotiations so far have failed to produce any agreement, and you have been brought in as an outside arbitrator. How can you propose a fair settlement of these issues? Indeed, what would "fair" mean in a context like this?

In this module, we will develop a classical arbitration scheme due to John Nash (1950) which gives one solution to this kind of problem.

Utility Theory

We will eventually address Management and Labor's problem, but to develop the theory, let's consider a much simpler example of arbitration. Suppose that Ellen and Frank are trying to decide how to spend their 16-day summer vacation. They are considering three alternatives: going to the mountains (M), going to the beach (B), or staying home (SQ). The SQ stands for "status quo," meaning that if they can't reach agreement, they'll stay where they are, i.e. stay home. They have different preferences among these alternatives, and they have called upon you, as a friend, to arbitrate.

The first thing you need to do is to find out how Ellen and Frank feel about the alternatives. You ask each of them for their *preference ordering*, and get

Ellen:	M	SQ	B
Frank:	B	SQ	M

In other words, Ellen's first choice is going to the mountains, her last choice is going to the beach, and Frank feels exactly the opposite.

To go further, you must find out something about the strength of Ellen's and Frank's preferences. Most economists agree that it is not useful to try to compare the strength of Ellen's preferences to the strength of Frank's. In other words, they refuse to become involved in arguments of the type: "I want the mountains more than you want the beach!" "No, I want the beach more than you want the mountains!" However, mathematician John von Neumann and economist Oskar Morgenstern (1944) suggested that it can be meaningful to ask about the relative strength of one person's preferences among the different alternatives. Their suggestion is the foundation of modern *utility theory.*

Here is the idea. Suppose Ellen can assign to each alternative a number in such a way that

- her more preferred alternatives get higher numbers
- the gaps between the numbers correspond to the relative strengths of her preferences.

We call such numbers *utilities*, and use the notation u_E for "Ellen's utility function." For instance, we might have

$$u_E(B) = 6 \qquad u_E(SQ) = 7 \qquad u_E(M) = 10.$$

We interpret this as saying that Ellen's preference for the mountains over staying home is three times as strong as her preference for staying home over the beach, since

$$u_E(M) - u_E(SQ) = 3\,(u_E(SQ) - u_E(B)).$$

But what does this really mean? Von Neumann and Morgenstern suggested a clever operational interpretation: Ellen would have no preference (would be *indifferent*) between

(i) getting SQ for certain, or

(ii) leaving the outcome to a chance device which chooses either B or M, with B three times as likely as M (i.e. the respective probabilities of choosing B and choosing M are 3/4 and 1/4).

The idea is that Ellen's preference for M over SQ, being three times as strong as her preference for SQ over B, would be exactly balanced by B being three times as likely to be chosen as M in (ii).

Von Neumann and Morgenstern called the kind of situation in (ii) a *lottery*, and built the entire theory of utility around lotteries. Davis (1970) gives a nicely accessible discussion of their theory. For our purposes, we will assume that Ellen and Frank, and Management and Labor, can give us utilities which can be interpreted in terms of behavior in lottery situations.

There is a point implicit in the above discussion which will be crucial to arbitration theory. Suppose we add a constant b to all of Ellen's utilities, to get a new utility function $v_E = u_E + b$. Since all of the ratios of gaps between the utility differences remain the same, for example

$$v_E(M) - v_E(SQ) = 3(v_E(SQ) - v_E(B)),$$

v_E gives exactly the same information about Ellen's behavior in lotteries as u_E. The same thing happens if we multiply a utility function by a positive constant. We are led to the following

Definition. *Two utility functions u and v are equivalent if there are constants $a > 0$ and b such that $v = au + b$.*

For example, all of the following functions give exactly the same information about Ellen's preferences:

$$\begin{array}{lll} u_E(B) = 6 & u_E(SQ) = 7 & u_E(M) = 10 \\ v_E(B) = -1 & v_E(SQ) = 0 & v_E(M) = 3 \\ w_E(B) = -100 & w_E(SQ) = 0 & w_E(M) = 300. \end{array}$$

Mathematicians say utility functions are ***linearly invariant.***

It is now becoming clear why we can't compare two different individuals' strengths of preferences. Suppose

$$\begin{array}{lll} u_E(B) = 6 & u_E(SQ) = 7 & u_E(M) = 10 \\ u_F(B) = 10 & u_F(SQ) = 3 & u_F(M) = 2. \end{array} \tag{1}$$

Can we conclude that Frank's preference for the beach over the mountains is stronger than Ellen's preference for the reverse, because $10 - 2 > 10 - 6$? Certainly not, since exactly the same preference information would be conveyed if we multiplied Ellen's utilities by 10.

For the same reason, it does not make sense to add two different individuals' utilities. For instance, could we conclude that Ellen and Frank should go to the beach because B has the largest sum of utilities $(6 + 10 > 10 + 2 > 7 + 3)$? If you think so, try multiplying Ellen's utilities by 10 and redoing the calculation.

Exercise

1. Which of the following utility functions is equivalent to u_F in (1)? For the one which is, find the a and b which give the equivalence.

$$\begin{array}{lll} v_F(B) = 10 & v_F(SQ) = -4 & v_F(M) = -6 \\ w_F(B) = 20 & w_F(SQ) = 2 & w_F(M) = -1 \end{array}$$

The Payoff Polygon

We will take advantage of the linear invariance of utility to add an appropriate constant to each of our utility functions to make $u(SQ) = 0$. This is called *normalizing* the utility functions. Thus for Ellen and Frank we will use the utility functions

$$\begin{array}{lll} u_E(B) = -1 & u_E(SQ) = 0 & u_E(M) = 3 \\ u_F(B) = 7 & u_F(SQ) = 0 & u_F(M) = -1 \end{array} \tag{1'}$$

One way to picture this information is to plot the alternatives B, SQ and M on a coordinate plane, with Ellen's utility on the horizontal axis and Frank's utility on the vertical axis:

$$B = (-1, 7) \qquad SQ = (0, 0) \qquad M = (3, -1).$$

See Figure 1a. Our job would seem to be to choose one of those three points as the fairest outcome. However, if those were the only choices, we would have no chance to propose a compromise. The von Neumann-Morgenstern theory poses a way out: we can propose a lottery as a solution. The utility of a lottery is easy to calculate.

Definition. *Suppose a lottery L is between outcomes P and Q, with probability p of P and probability $1 - p$ of Q. Then for any utility function u, the utility of the lottery is $u(L) = pu(P) + (1 - p)u(Q)$.*

This follows directly from von Neumann and Morgenstern's interpretation of utility in terms of indifference between lotteries. In our coordinate plane representation for Ellen and Frank, it has a nice interpretation. Suppose we propose the lottery L with probability 3/4 of B, 1/4 of M. Then

$$u_E(L) = \frac{3}{4}(-1) + \frac{1}{4}(3) = 0 \quad \text{and} \quad u_F(L) = \frac{3}{4}(7) + \frac{1}{4}(-1) = 5,$$

so in the coordinate plane $L = (0, 5)$. Geometrically, this point is on the line segment between B and M, 3/4 of the way toward B. A little thought should convince you that this result generalizes to any lottery between a pair of outcomes: the point corresponding to it will lie on the line segment joining those outcomes, with the probabilities determining the exact location. In Exercise 2 you are asked to consider a lottery among three outcomes. You'll find that the corresponding point is inside the triangle determined by those outcomes.

The result of this analysis is that if we are allowed to propose lotteries as outcomes, we can propose any point in or on the boundary of the triangle M-SQ-B. This triangle is called the ***payoff polygon*** for the problem. See Figure 1b.

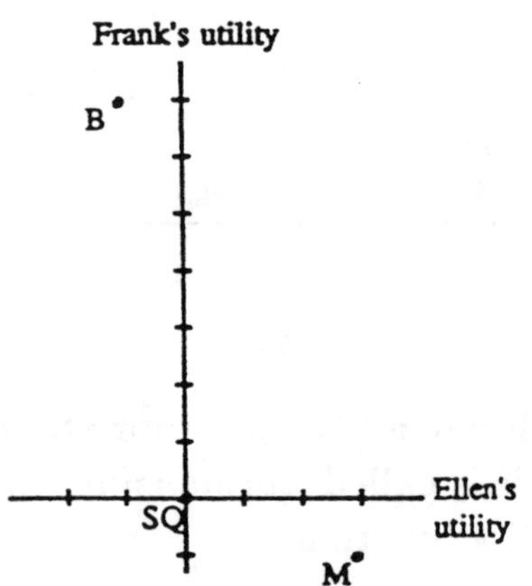

a. Ellen and Frank's outcome points.

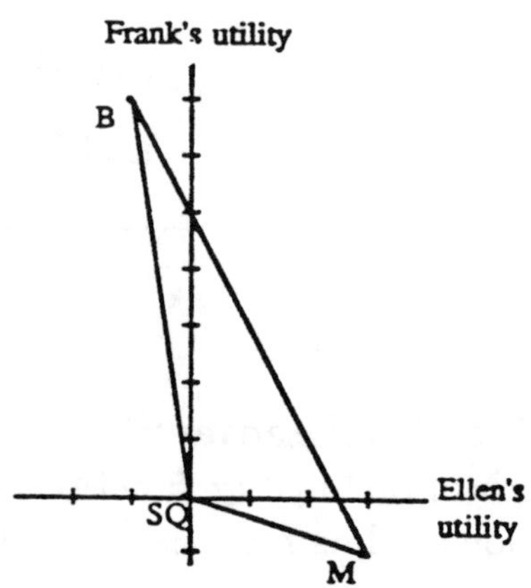

b. Ellen and Frank's payoff polygon.

Figure 1.

Definition. *The payoff polygon for an arbitration problem with possible outcomes A, B, ..., SQ is the smallest convex polygon containing all of these points.*

The polygon in the definition is sometimes called the ***convex hull*** of the outcome points. It's what you would get if you put pegs at each of the outcome points and let a big rubber band close around the outside. Figure 2 shows an example.

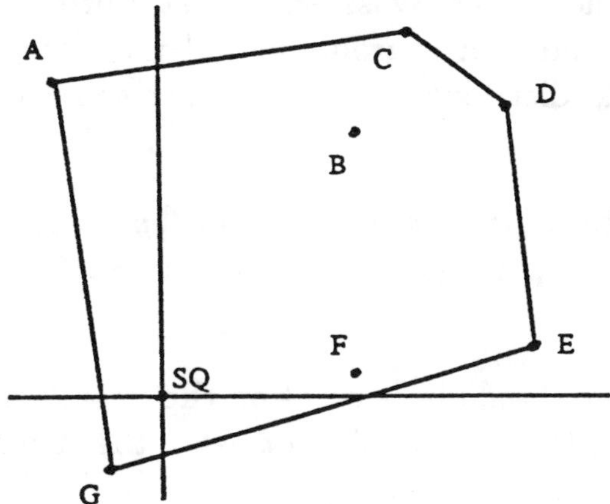

Figure 2. A payoff polygon for 8 points.

Exercises

2. a) In Figure 1, find the coordinates of the points corresponding to the following lotteries:

   L_1 : SQ with probability 1/3, M with probability 2/3.

   L_2 : SQ with probability 1/2, M with probability 1/4, B with probability 1/4.

   b) What lottery corresponds to the point (2.5,0) in Figure 1?

3. Draw, carefully on graph paper, the payoff polygon for the arbitration problem with outcomes

$$A = (4,2), B = (2,6), C = (0,7), D = (1,5), E = (-4,2), F = (0,-2), SQ = (0,0).$$

The Nash Arbitration Scheme

Of course, we haven't solved our arbitration problem yet, but we now have a geometric picture of what we need to do. An arbitration situation can be modeled, as we have done, as a convex polygon containing a status quo point. We have agreed to choose utility scales for our agents so that the status quo point is (0,0). We must propose a particular point in the polygon as a solution. We would like to have a method which tells us how to do this systematically for any arbitration situation. The following definition is due to John Nash (1950).

Definition. *An arbitration method is a rule which assigns to each convex polygon $\mathcal{P}$ containing the origin (0,0), a unique point $N = (x^*, y^*)$ in $\mathcal{P}$.*

Of course, we want to find not just any arbitration method, but one which has some claim to being "good" and "fair." Nash's approach was to begin by writing down conditions which a good arbitration scheme should satisfy. The first condition asks for a kind of acceptability introduced by the Italian-Swiss economist Vilfredo Pareto at the beginning of the 20th century. Pareto said that a group should not accept a given distribution of wealth if some other possible distribution would make some people better off without making anyone worse off.

Condition 1 (Pareto). *The solution point $N = (x^*, y^*)$ should be Pareto acceptable. That is, there should not be any point (x, y) in $\mathcal{P}$ with $x > x^*$ and $y \geq y^*$, or with $x \geq x^*$ and $y > y^*$.*

Geometrically, this says that N should be a point on the "northeast" part of the boundary of $\mathcal{P}$. The next condition says that we should not ask either agent to accept less than he or she has at the status quo.

Condition 2 (Rationality). *The solution point $N = (x^*, y^*)$ should have $x^* \geq 0$ and $y^* \geq 0$.*

The third condition acknowledges that either or both agents' utility scales may be multiplied by any positive constant, and this should not change the solution.

Condition 3 (Linear Invariance). *Suppose that a and b are positive constants, and we consider the transformation $T(x, y) = (ax, by)$, which multiplies x-coordinates by a and y-coordinates by b. If the N is the solution for $\mathcal{P}$, then the solution for $T(\mathcal{P})$ should be $T(N)$.*

See Figure 3a, which shows an example with $a > 1$ and $0 < b < 1$. Condition four embodies the most elementary notion of fairness:

Condition 4 (Symmetry). *If $\mathcal{P}$ is symmetric across the line $x = y$, then N should be on this line.*

The last condition is a little more involved.

Condition 5 (Independence of Irrelevant Alternatives). *Suppose that N is the solution for $\mathcal{Q}$, and $\mathcal{P}$ is a polygon which is completely contained in $\mathcal{Q}$, and contains both $(0, 0)$ and N. Then N should also be the solution for $\mathcal{P}$.*

See Figure 3b. The idea here is to suppose that our parties, when confronted with all of the possible outcomes in $\mathcal{Q}$, have agreed that N is the fairest outcome. Then they discover that not all of the outcomes in $\mathcal{Q}$ were really available—only those in $\mathcal{P}$ were. They should still agree that N is fairest. Their perceptions should not have been affected by those "irrelevant" outcomes which are in $\mathcal{Q}$ but not in $\mathcal{P}$.

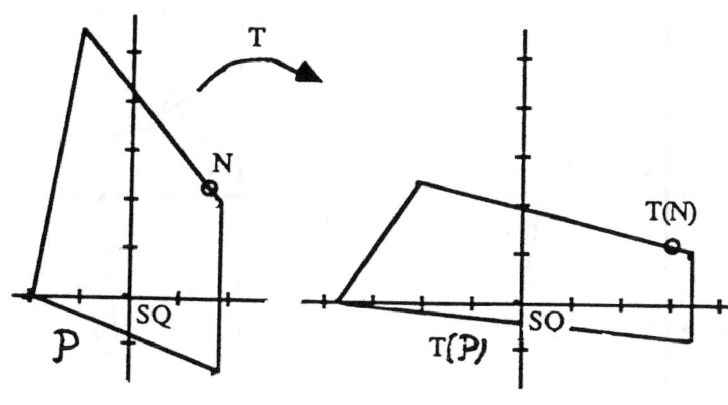

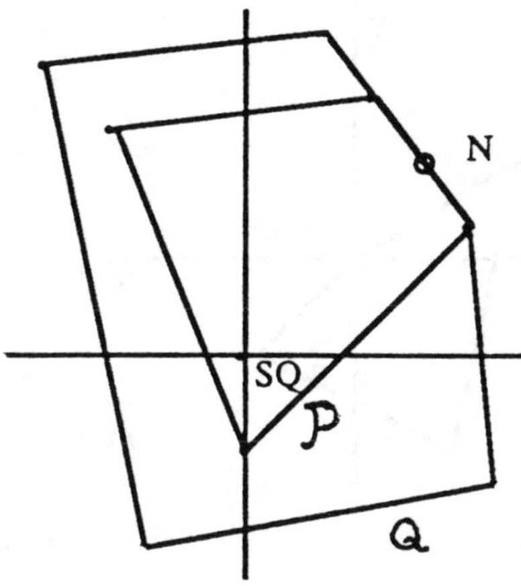

a. Linear Invariance **b. Independence of Irrelevant Alternatives**

Figure 3.

I hope you agree that these conditions all do seem reasonable. If you do, there is a wonderful surprise. Nash proved that there is <u>one and only one</u> arbitration method which satisfies these axioms. Moreover, it is easy to compute, using just a little calculus.

Theorem (Nash, 1950). *There is exactly one arbitration which satisfies Conditions 1 through 5. It is the following:*

i) If there are no points in $\mathcal{P}$ with $x > 0$ or $y > 0$, let $N = (0,0)$, the status quo.

ii) If there are no points in $\mathcal{P}$ with $y > 0$, but are points with $y = 0$ and $x > 0$, then let N be the point $(x, 0)$ which maximizes x. Handle the case with x and y interchanged similarly.

iii) If there are points in $\mathcal{P}$ with both $x > 0$ and $y > 0$, let N be the point with $x > 0$ and $y > 0$ which maximizes the <u>product</u> xy.

Figure 4 shows pictures of the three cases. The first two cases are no surprise: the solution is just what is required by Conditions 1 and 2. The third, and most useful, case is surprising, and it is the one we will work on proving and calculating in examples. We have seen that maximizing the sum of utilities wouldn't make sense in the modern interpretation of utility. Nash says maximize the <u>product</u> of utilities. However, this is not because the product has some *a priori* virtue, but because maximizing the product is the only way we can satisfy Nash's set of reasonable conditions. Doing the maximization is, of course, where calculus comes in.

The proof of Nash's Theorem involves just some elementary geometry, but we'll put it off until the end. Right now, let's use it to solve some arbitration problems.

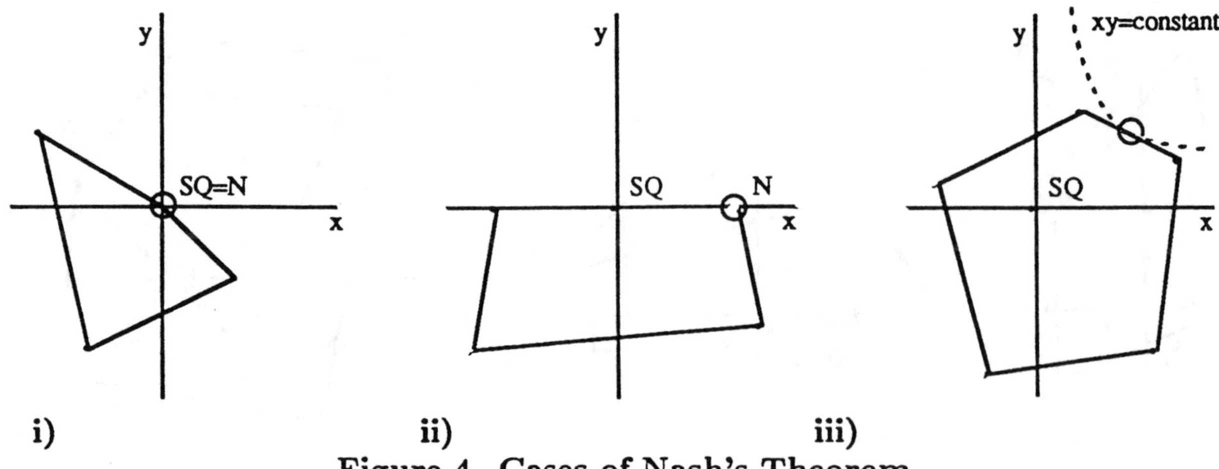

Figure 4. Cases of Nash's Theorem.

Calculations

To get a feel for how the Nash method works in practice, we will work through some examples.

I. Let's start with Ellen and Frank's problem, as shown in Figure 1b. The solution point will be in the first quadrant on the northeast boundary, which is on the line $y = 5 - 2x$. Our problem is

$$\text{Maximize } xy \quad \text{subject to} \quad y = 5 - 2x \quad \text{on the interval} \quad 0 \leq x \leq 5/2.$$

Substituting for y gives the problem

$$\text{Maximize } f(x) = x(5 - 2x) = 5x - 2x^2 \quad \text{on } [0, 5/2].$$

Calculus tells us that the maximum must occur at either an endpoint of the interval, $x = 0$ or $x = 5/2$, or at a *critical point* in the interval, where $f'(x) = 0$. Solving $f'(x) = 5 - 4x = 0$ gives a unique critical point $x = 5/4$, which is in the interval $[0, 5/2]$. We evaluate $f(x)$ at these points and get

Type of point	x	$f(x)$
endpoint	0	0
endpoint	5/2	0
critical point	5/4	25/8

The maximum is at $x = 5/4$. When $x = 5/4$, $y = 5 - 2x = 5/2$. Thus $N = (5/4, 5/2)$, as in Figure 5a. Of course, we need to be able to tell Ellen and Frank what this means. It is a point on the line segment between M and B, so corresponds to a lottery involving M and B. Because the utility to Ellen is 5/4, we can find the probabilities by solving

$$5/4 = p(-1) + (1 - p)(3) \qquad \text{(using the } x\text{-coordinates of } N, M, B)$$

for p, which gives $p = 7/16$. So $N = (7/16)B + (9/16)M$. We might propose that Ellen and Frank put 7 red balls and 9 green balls in a jar and choose one ball at random: red means "go to the beach," and green means "go to the mountains."

In practice, proposing a lottery as a solution may not be very feasible, and we should try to use some flexibility and creativity in interpreting the Nash solution. For example, recalling that Ellen and Frank had 16 days of vacation, we might propose that they spend 7 days at the beach and 9 in the mountains.

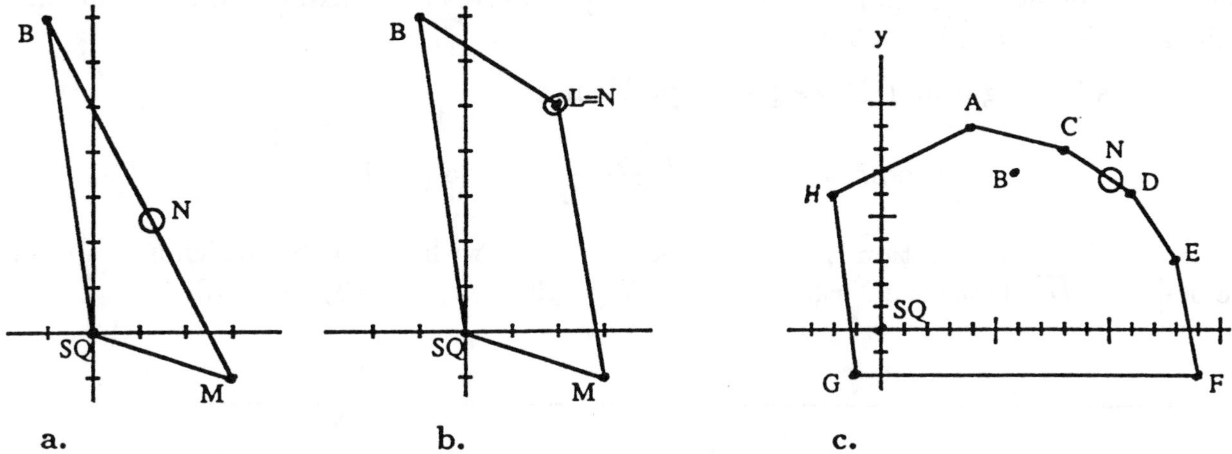

Figure 5. Payoff Polygons for Calculational Examples.

II. Suppose Ellen and Frank have another option: going to a lake (L). Assume that, with the normalized utility scales as above, L has coordinates $(2,5)$. See Figure 5b. N could now be on the line $y = 19/3 - (2/3)x$ with x in $[0,2]$, or on the line $y = 17 - 6x$ with x in $[2, 17/6]$. For the first line we get, by the procedure above, the problem

$$\text{Maximize } f(x) = x\left(\frac{19}{3} - \frac{2}{3}x\right) \quad \text{on } [0,2].$$

The only critical point $x = 19/4$ is not in the interval [0,2]. The maximum on the interval is at the endpoint $x = 2$. You should check that the second line segment gives a critical point $x = 17/12$ which is outside of [2,17/6], and the maximum is at the endpoint $x = 2$. We conclude that N is the corner point (2,5). Ellen and Frank should go to the lake.

III. Let's consider a more abstract problem, with alternatives

$$A = (4,9),\ B = (6,7),\ C = (8,8),\ D = (11,6),\ E = (13,3),$$

$$F = (14,-2),\ G = (-1,-2),\ H = (-2,6),\ SQ = (0,0).$$

Here it is really important to draw the payoff polygon, as in Figure 5c. We see, for instance, that B is not Pareto acceptable, and that the solution could be on any of the line segments $\overline{AC}$, $\overline{CD}$, $\overline{DE}$ or part of $\overline{EF}$. We need not check them all. Compute the products of coordinates at each corner:

$$A:\ 4\cdot 9 = 36 \qquad C:\ 8\cdot 8 = 64 \qquad D:\ 11\cdot 6 = 66 \qquad E:\ 13\cdot 3 = 39$$

A little thought should convince you that the maximum must be on a line segment adjacent to the highest corner value, which is D. So we only need to check $\overline{CD}$ and $\overline{DE}$. (Unfortunately, we cannot conclude that the maximum must be on $\overline{CD}$ just because C has the second highest product of coordinates.) Checking $\overline{CD}$ first, the problem becomes

$$\text{Maximize } f(x) = x\left(\frac{40}{3} - \frac{2}{3}x\right) \quad \text{on } [8, 11].$$

The critical point $x = 10$ is in the interval $[8, 11]$ and gives the maximum on the interval, with value $f(10) = 200/3 = 66\frac{2}{3}$.

For the line segment $\overline{DE}$, we get the problem

$$\text{Maximize } g(x) = x\left(\frac{45}{2} - \frac{3}{2}x\right) \quad \text{on } [11, 13].$$

For this problem the solution is at the endpoint $x = 11$, with $g(x) = 66$, which is less than the $66\frac{2}{3}$ on $\overline{CD}$. Hence the Nash solution is $N = (10, 20/3) = (1/3)C + (2/3)D$.

Exercises

4. Use the Nash arbitration scheme to solve the following arbitration problems. If the solution is not one of the given alternatives, express it as a lottery combination of two of the given alternatives.

   a. The arbitration problem in Exercise 3.

   b. $A = (5, 26)$, $B = (12, 21)$, $C = (14, 19)$, $D = (16, 17)$, $E = (20, 7)$, $F = (-5, 0)$, $G = (0, -3)$, $SQ = (0, 0)$.

5. The English philosopher R.B. Braithwaite (1955) imagined the following situation: Luke and Matthew are both bachelors and occupy adjacent apartments with a very thin wall between them. They each have just one hour a day—between 9 and 10 p.m.—available for recreation. Luke likes to play classical music on his piano, while Matthew enjoys improvising jazz on a trumpet. Currently, they both play, but this is unpleasant for both. Can you suggest a fair compromise?

   a. The choices are Luke plays alone (L), Matthew plays alone (M), both play (SQ), or neither play (N). Utility questions elicit

$$u_L(SQ) = 0 \quad u_L(N) = 5 \quad u_L(M) = 6 \quad u_L(L) = 10$$

$$u_M(N) = 0 \quad u_M(SQ) = 3 \quad u_M(L) = 7 \quad u_M(M) = 10.$$

   What do you recommend? [Note that Matthew's utilities as given are not yet normalized.]

b. Suppose everything is the same, except that in Matthew's preferences N and SQ trade places (so that Matthew, like Luke, now prefers silence to cacophony).

What do you recommend now? Think about and comment on these results.

6. Kalai and Smorodinsky (1975) considered the following arbitration situations:

$$i)\quad A = (0,1),\ B = (1,0),\ C = (.75,.75),\ SQ = (0,0).$$

$$ii)\quad A = (0,1),\quad B = (1,0),\quad C = (1,.7),\quad SQ = (0,0).$$

Carefully draw the payoff polygons for these situations on the same axes. Explain why the person whose utilities are on the y-axis should expect to do better in situation ii) than in situation i). Then solve the two problems—no calculation should be necessary for i)—and compare the results for the person whose utility is on the y-axis. Kalai and Smorodinsky were upset enough by the results to propose an alternative to Nash's scheme.

Solution to the Management-Union Arbitration Problem

We'll apply the Nash arbitration scheme to Management and Labor's problem from the first section. Recall that the individual items under discussion are

A: automation of the checkpoint
C: elimination of the coffee break
P: the pension benefit package
R: the one dollar raise
SQ: the status quo (no change from the present)

The outcome of the arbitration will either be SQ, or some combination of one or more of the items. For example, one possible outcome could be denoted ACR: Labor agrees to automating the checkpoint and eliminating the coffee break, and Management gives the raise. There are 16 possible outcomes:

$$SQ, A, C, P, R, AC, AP, AR, CP, CR, PR, ACP, ACR, APR, CPR, ACPR.$$

It would be complicated to ask Management and Labor for their utilities for 16 outcomes. Fortunately, a simplification is possible if Management, say, values the items A, C, P and R *independently*. This means that the value of any item is independent of whether some other item is obtained: giving the raise costs Management just as much, regardless of whether or not the checkpoint is automated. In this case, Management's utilities will be *additive*, in the sense that

$$u_M(AR) = u_M(A) + u_M(R), \qquad \text{and} \qquad u_M(ACR) = u_M(A) + u_M(C) + u_M(R).$$

For simplicity of the analysis, we will assume that both Management and Labor have additive utilities. This is not completely unreasonable. When my student Phil Polgreen (1991) asked the management of a division of Sundstrand Corporation to give their utilities for issues in a labor dispute, they had an accounting firm's cost estimate for every issue—the bottom line was money, which is additive. Labor had a bit more trouble, but still gave utilities which were close to additive.

If utilities are additive, we need only determine utilities for the basic items. Suppose we find that they are (normalized):

$$u_M(R) = -3 \quad u_M(P) = -2 \quad u_M(SQ) = 0 \quad u_M(A) = u_M(C) = 4, \tag{2}$$

$$u_L(A) = -2 \qquad u_L(C) = -1 \qquad u_L(SQ) = 0 \qquad u_L(P) = 2 \quad u_L(R) = 3. \tag{3}$$

Thus Labor, for example, wants the raise more than the pension increase, and dislikes automating the checkpoint more than giving up the coffee break. They would consider an outcome of ACR as no better or worse than the status quo:

$$u_L(ACR) = u_L(A) + u_L(C) + u_L(R) = (-2) + (-1) + 3 = 0 = u_L(SQ).$$

Using the additivity of utilities, we can now calculate Management and Labor utilities for all 16 possible outcomes, plot them in the plane, and draw the payoff polygon in Figure 6.

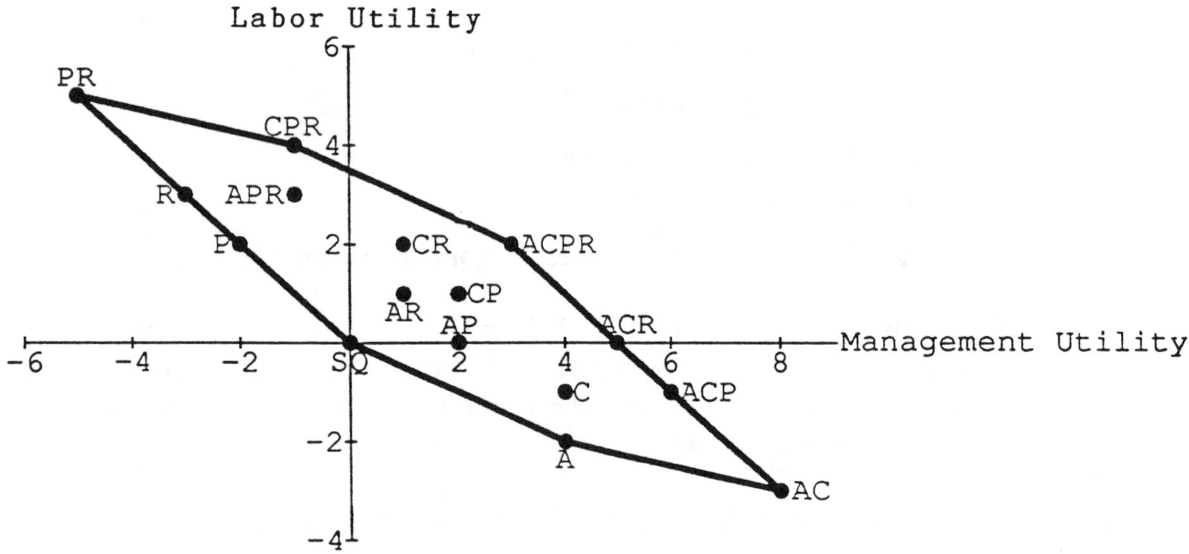

Figure 6. Management-Labor payoff polygon.

From the payoff polygon, we see that the solution must be on

$$CPR\text{-}ACPR: \quad \text{Maximize } f(x) = x\left(\frac{7}{2} - \frac{1}{2}x\right) \quad \text{on } [0,3]$$

$$\text{or} \quad ACPR\text{-}ACR: \quad \text{Maximize } g(x) = x\,(5 - x) \quad \text{on } [3,5].$$

The function $f(x)$ has its maximum at the endpoint $x = 3$, and $g(x)$ also has its maximum at the endpoint $x = 3$. We conclude that $N = (3, 2) = ACPR$. You as arbitrator, if you use this method, should propose that each side yield both concessions asked by the other side. The trade-off benefits both sides, and it is fair if we agree that Nash's axioms capture the idea of fairness.

Exercises

7. Carefully do the calculations to show that $N = (3, 2)$, as claimed above.
8. What would be the effect on the payoff polygon if $u_M(R) = -1$, i.e. Management really didn't mind very much giving the raise? Compute the new Nash outcome. Are there any ways this outcome might be implemented without a lottery?

Proof of Nash's Theorem

It is not hard to see that Nash's method does satisfy Conditions 1 through 5. Conditions 1 and 2 are easy. Condition 3 holds because if we have maximized xy, we have also maximized $abxy$. Condition 5 holds because points in $\mathcal{Q}$ but not in $\mathcal{P}$ must have smaller products of coordinates than N. For Condition 4, see Exercise 9.

The more surprising part of the theorem says that <u>any</u> arbitration method which satisfies Conditions 1 through 5 <u>must</u> be Nash's method. To show this, consider any arbitration polygon $\mathcal{P}$, which we will assume has points with $x > 0$ and $y > 0$ (so we are in case iii)). Let N be the Nash point, which maximizes xy. We must show that any arbitration method satisfying the Conditions must give N as its solution point for $\mathcal{P}$.

First of all, we can use Condition 3 to change utility scales to move N to the point $(1, 1)$—if N had been at (x^*, y^*), just multiply the utility scales by $a = 1/x^*$ and $b = 1/y^*$. So we can assume that for all other points in $\mathcal{P}$ with $x > 0$ and $y > 0$, $xy \leq 1$. Geometrically, this means that $\mathcal{P}$ lies entirely on or below the hyperbola $xy = 1$, and it touches the hyperbola at $(1, 1)$. Since $\mathcal{P}$ is a convex polygon, it follows that $\mathcal{P}$ must lie entirely on or below the tangent line to the hyperbola at $(1, 1)$, which is the line $x + y = 2$. See Figure 7.

The rest of the proof is beautifully simple. Since $\mathcal{P}$ lies on or below $x + y = 2$, enclose $\mathcal{P}$ in a large rectangle $\mathcal{Q}$ which has one side on this line and is symmetric about the line $x = y$, as shown in Figure 7. By Conditions 1 and 4, the solution for $\mathcal{Q}$ must be at $(1, 1)$. Then by Condition 5, the solution for $\mathcal{P}$ must also be at $(1, 1)$, and we are done!

That is one of the most elegant proofs I know. Notice the power of Condition 5, which finishes the proof dramatically.

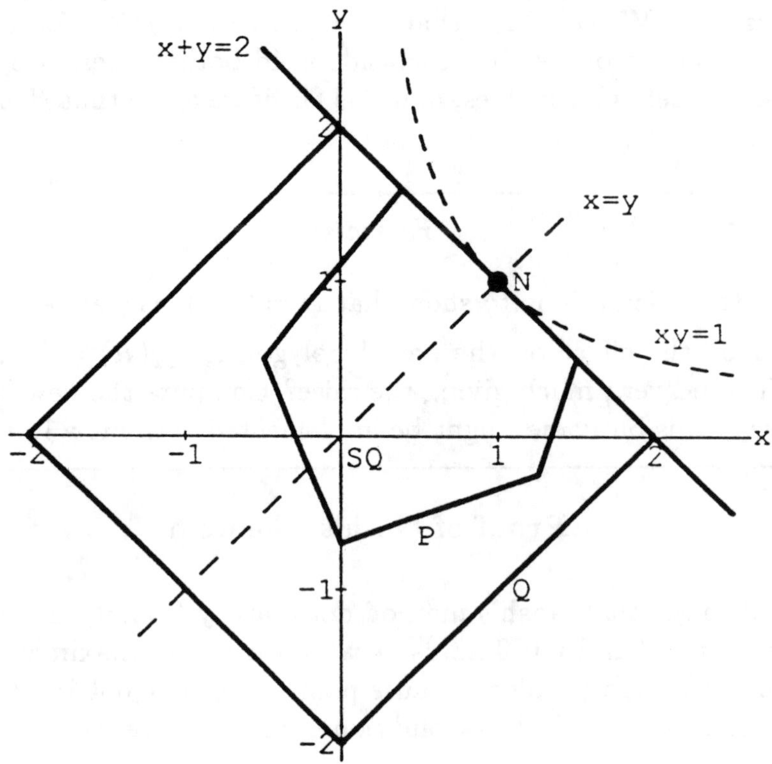

Figure 7. Proof of Nash's Theorem.

Exercise

9. Let's show that Nash's method satisfies Condition 4. If $\mathcal{P}$ is symmetric, then whenever a point (a, b) is in $\mathcal{P}$, so is (b, a). Since $\mathcal{P}$ is convex, the whole line segment from (a, b) to (b, a) must also be in $\mathcal{P}$. Show that on this line segment, the maximum value of xy occurs at $(\frac{a+b}{2}, \frac{a+b}{2})$, which is on the line $x = y$.

Further Reading

The suggestion that Nash's arbitration theory could be applied to labor negotiations was first made in (Allen, 1956), which contains a number of nicely worked out examples. (Polgreen, 1992) applies the theory to an actual labor negotiation in California in 1989. There are certainly some practical problems. For instance, it may be difficult to obtain utilities from the parties involved. Another problem is honesty: there is no reason to expect that a party will answer preference questions honestly if he sees that dishonest answers could produce an arbitrated solution more favorable to him. I have argued in (Straffin, 1993) that lying is a tricky business in the Nash scheme. The obvious kinds of lies don't always produce an advantage, and when both parties lie, there is a good chance that both parties will end up worse off than if they had both told the truth.

While we're talking about nasty behavior, there is the matter of threats. Not all labor negotiations are amicable: Labor often mentions strikes, and Management sometimes talks of lockouts. We can think of threats as attempts to move the status quo point downward (Management threats) and to the left (Labor threats). When we consider how to "threaten optimally," we enter the fascinating area of mathematical game theory—see below for general references. Nash presented his idea on optimal threats in (Nash, 1953).

On the other hand, thinking of arbitration as a cooperative endeavor, as the Nash scheme does, has produced some ideas which are now widely used in arbitration. One is to encourage both sides to make as many "demands", and as many offers, as possible. The more items are under consideration, the bigger the polygon $\mathcal{P}$ is likely to be, and the more likely it is that there will be combinations of items which will give points in the first quadrant—deals beneficial to both sides.

There have been several specific criticisms of the fairness of the Nash scheme. One was based on the results of Exercise 6, which lead to the proposal of an alternate arbitration scheme in (Kalai and Smorodinsky, 1975). There is a very thoughtful and thorough discussion of fairness and Nash's conditions in Chapter 6 of (Luce and Raiffa, 1957). Chapter 16 of (Raiffa, 1982) discusses both the fairness and the practicality of Nash's scheme in the form of an imaginary dialogue between a practical Arbitrator and a mathematical Analyst.

References

Allen, Layman (1956), "Games bargaining: a proposed application of the theory of games to collective bargaining," *Yale Law Journal* 65: 660-693.

Braithwaite, R.B. (1955), *Theory of Games as a Tool for the Moral Philosopher*, Cambridge University Press.

Davis, Morton (1970), *Game Theory: A Non-Technical Introduction*, Basic Books.

Kalai, E. and M. Smorodinsky (1975), "Other solutions to Nash's bargaining problem," *Econometrica* 43: 513-518.

Luce, R.D. and Howard Raiffa (1957), *Games and Decisions*, John Wiley and Sons.

Nash, John (1950), "The bargaining problem," *Econometrica* 18: 155-162.

Nash, John (1953), "Two-person cooperative games," *Econometrica* 21: 128-140.

Owen, Guillermo (1982), *Game Theory*, second edition, Academic Press.

Polgreen, Philip (1992), "Nash's arbitration scheme applied to a labor dispute," *The UMAP Journal* 13: 25-35.

Raiffa, Howard (1982), *The Art and Science of Negotiation*, Harvard University Press.

Straffin, Philip (1993), *Game Theory and Strategy*, Mathematical Association of America.

Von Neumann, John and Oskar Morgenstern (1944), *Theory of Games and Economic Behavior*, Princeton University Press. Third edition by John Wiley and Sons, 1967.

Answers to Exercises

1. $v_F = 2u_F - 10$.
2. a) $L_1 = (1/3)(0,0) + (2/3)(3,-1) = (2, -(2/3))$; $L_2 = (1/2)(0,0) + (1/4)(3,-1) + (1/4)(-1,7) = (1/2, 3/2)$.

   b) $(2.5, 0) = (7/8)M + (1/8)B$.
3. D and SQ are in the interior of the pentagon $ABCEF$.
4. a) $N = (2.5, 5) = (1/4)A + (3/4)B$.

   b) $N = (16, 17) = D$.
5. a) $N = (7/12)M + (5/12)L$. Matthew could play for 35 minutes, Luke for 25.

   b) Now $N = (1/12)M + (11/12)L$, and Matthew gets only 5 minutes. Most of Matthew's bargaining strength came from the fact that he preferred cacophony to silence.
6. Symmetry and Pareto acceptability say that the solution to a) must be at C. The solution to b) also turns out to be at C, and y is less than in a). The top boundary of the polygon has been moved up—for every value of x there are larger values of y available—and yet this has hurt the vertical party. Kalai and Smorodinsky think that this is unfair.
8. If $u_M(R)$ increases from -3 to -1, all of the points involving R will move 2 units to the right. The new Nash outcome will still be on the line segment between $CPR = (1,4)$ and $ACPR = (5,2)$, or on the segment between $ACPR$ and $ACR = (7,0)$. It works out that $N = (4.5, 2.25) = (7/8)ACPR + (1/8)CPR$. This time, the union should give up only 7/8 of A. If you remember that A would cost eight union jobs, perhaps management could hire one union member back to supervise the checkpoint!
9. The equation of the line segment from (a,b) to (b,a) is $y = a + b - x$. The maximum of xy along this line segment between $x = a$ and $x = b$ is indeed at $x = (a+b)/2$.

FITTING LINES TO DATA

Author: Thomas L. Moore, Grinnell College, Grinnell, IA 50112

Calculus Needed: Derivatives to find local extrema, logarithm and exponential functions.

Area of Application: Statistics

Comments: Students should use computer software or a graphics calculator to do the calculations and make scatterplots for exercises marked $*$. Pro-Matlab was used to make the plots in this document. Other exercises can be done by hand or calculator. Data sets not in the main text are in the Appendix.

A Problem: Estimating the Amount of Lumber in a Woodlot

A forester wishes to estimate the amount of black cherry tree timber in woodlots in the Allegheny National Forest. By choosing a sample of trees of a given species on a woodlot and measuring the volume of wood in each tree the forester can estimate the total volume of lumber for that species in that woodlot.

Unfortunately it is difficult to measure the volume of a tree. Imagine trying to do so. Could you do it without cutting down the tree? Since the tree is very irregularly shaped, many measurements would be required to calculate its volume accurately. Measuring the volume of a single tree would require much time and effort and this process would be required for every tree in the sample.

A different strategy for solving the problem is to use a procedure called *regression analysis*. The data in Table 1 give diameter, height, and volume measurements on a representative sample of 31 black cherry trees. The diameter is the diameter in inches measured at a height of 4.5 feet above ground level (the so-called breast-height diameter). The height is in feet and the volume in cubic feet. Someone worked hard to figure the volumes of these 31 trees with the hope that through regression analysis we wouldn't have to figure the volumes of any other trees.

While it is difficult to measure the volume of a tree in the field, measuring its height or its diameter is much easier. We can imagine measuring the diameter readily in a few seconds. The height is not as accessible, but with the help of some right triangle geometry it can be measured with fair accuracy. In fact, people measuring woodlots—"cruising timber" as it is called—often use a *cruising stick* that allows quick measurements of both diameter and height.

Diameter	Height	Volume
8.3	70	10.3
8.6	65	10.3
8.8	63	10.2
10.5	72	16.4
10.7	81	18.8
10.8	83	19.7
11.0	66	15.6
11.0	75	18.2
11.1	80	22.6
11.2	75	19.9
11.3	79	24.2
11.4	76	21.0
11.4	76	21.4
11.7	69	21.3
12.0	75	19.1
12.9	74	22.2
12.9	85	33.8
13.3	86	27.4
13.7	71	25.7
13.8	64	24.9
14.0	78	34.5
14.2	80	31.7
14.5	74	36.3
16.0	72	38.3
16.3	77	42.6
17.3	81	55.4
17.5	82	55.7
17.9	80	58.3
18.0	80	51.5
18.0	80	51.0
20.6	87	77.0

Table 1. Diameter, height, and weight for a sample of black cherry trees from the Allegheny National Forest, Pennsylvania. Source: [RJR], p. 328.

For the moment let's ignore the height measurements. Figure 1 shows a *scatterplot* of volume versus diameter for our 31 trees. It consists of each of the 31 ordered pairs (diameter,volume) plotted in x-y fashion. For example, the first tree in Table 1 is plotted as the point $(8.3, 10.3)$ in the scatterplot. Note how closely related volume is to diameter. Trees with small diameter have small volume and as we look at trees with increasingly larger diameter the volumes become progressively larger in a fairly predictable way. In fact, the relationship is nearly linear. That is, we can draw a line through the data that "fits" the data well as in Figure 1. The equation of the line shown is

$$\text{Volume} = -37 + 5 * \text{Diameter}.$$

Now if you give me the diameter of a black cherry tree and I use the equation to ***predict*** the volume, I will probably not be far off, provided that this sample of trees is representative of the forest. For example, if a tree has a diameter of 15 inches, we will estimate its volume to be

$$\text{Volume} = -37 + 5 * 15 = 38 \text{ cubic feet.}$$

Armed with this simple equation, the forester can now easily estimate the amount of lumber in his or her woodlot.

Our estimates will depend on which line we fit to the data. To find out how to fit lines to data, we need to investigate the area of statistics called regression analysis. We will find that this method can be a useful tool for analyzing not only linear relationships between variables, but non-linear ones as well.

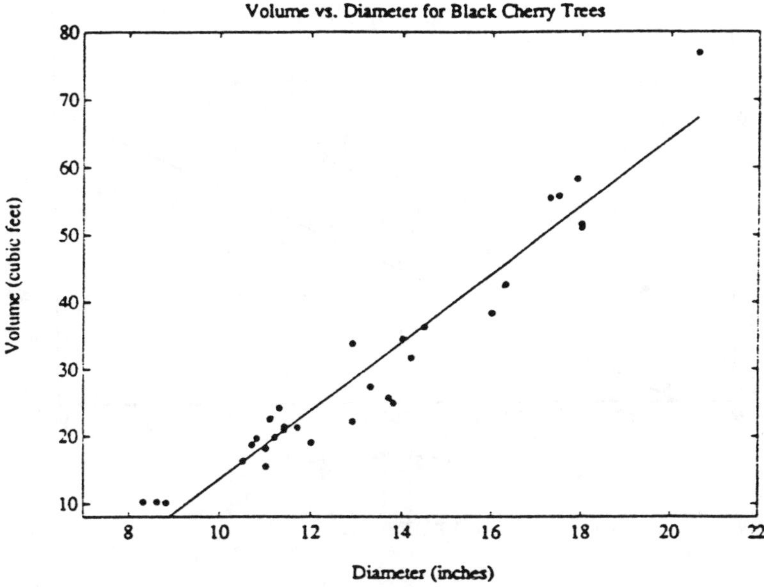

Figure 1.

Regression

Regression analysis is a collection of tools for exploring relationships between variables. In the timber example above, we have three variables (or measurements) made on each of the 31 trees in the sample. The variable of interest is volume, but volume is hard to measure. So we are interested in predicting volume from either or both of the more easily measureable variables, diameter and height.

Relationship between two variables refers to the degree to which values of one variable rise or fall with values of the second variable. For example, we say there is a positive relationship between the volume of a tree and its diameter. In our data set, as diameter increases so does volume. As indicated in section 1, we can imagine an underlying straight line relationship between volume and diameter. Because of extraneous factors beyond our control our data does not precisely follow a straight line. These "random" deviations off of our hypothesized straight line we call *statistical error* or simply *error*. The term *error* is conventional and should not be confused with error in the sense of a mistake. It simply means unknown and random deviations from some underlying model.

Figure 2 shows volume versus height for black cherry trees. Here again we have a positive relationship between y and x and again we can imagine the relationship as basically linear. But in this case the deviations of the actual data from the line are larger, so we say the relationship between volume and height is *weaker* than that between volume and diameter.

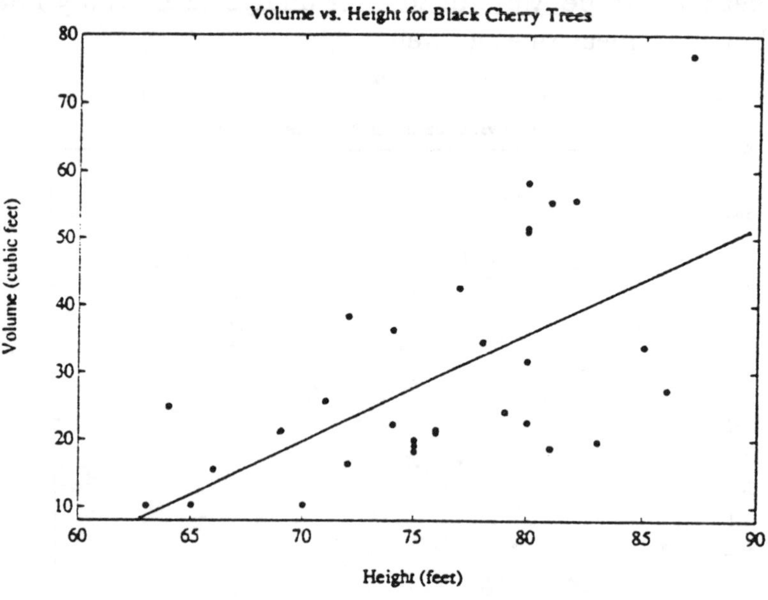

Figure 2.

Figure 3 shows the population size over time of a colony of mammary cancer cells. Here there is also a positive relationship between y (population size) and x (time) in that as time increases the population clearly increases. However, in this case, one would probably not consider a straight line as the best description of the underlying relationship. A better description would have to be a more complicated curve, perhaps an exponential curve.

In regression analysis we try to describe the relationship using a so-called *model*. A model is simply a mathematical description of the relationship between y and x. We usually write a model in the form

$$y = f(x) + error$$

where $f(x)$ is some simple mathematical function (e.g. a linear function for Figure 1 or an exponential function for Figure 3) and *error* represents random deviations from $f(x)$.

The choice for $f(x)$ may be guided by theoretical considerations. For example, we know from experience that exponentials are often good for describing population growth.

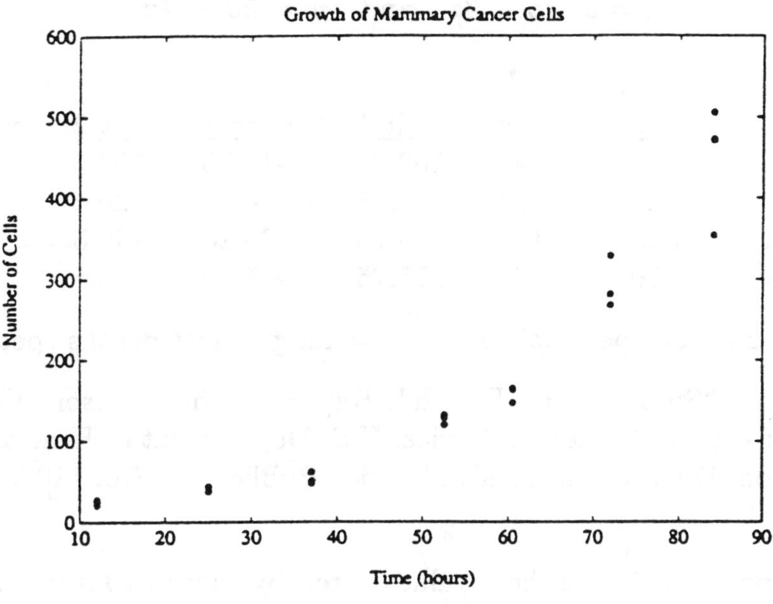

Figure 3.

In this module we will restrict our attention to the straight line model. Having seen Figure 3, you may feel that this is too narrow a focus, but in fact, the straight line is more powerful than it first appears for the following reasons:

- Many relationships are linear or nearly so.
- Even non-linear relationships are close to linear over a limited range of the x variable. In some cases, especially when the actual form of the non-linear relationship is unknown, a practical solution might be to fit a straight line to the range of x of primary interest, if that range is small.
- Transformations: a non-linear relationship can often be transformed to a linear relationship. We will see several examples at the end of this module.

Best Fit Lines

Given a scatterplot and a line through the plot we need a measure of fit in order to compare the fit of different lines to a given set of data. By a *measure of fit* we mean a single number that summarizes how well a given line fits a given scatterplot. Smaller values will indicate better fit. If the data fall exactly on a straight line, the measure of fit between the line and the data will be zero. Having chosen a measure of fit, we can use it to find the line that minimizes this measure for a given scatterplot and this will provide an objective algorithm for choosing the best fit line to a scatterplot.

Let's use a simple example to illustrate. Table 2 contains data for four (x, y) points. Two lines are drawn through the data:

$$y = 34.75 + .3x \quad \text{and} \quad y = .50 + .4x$$

	x	y	fit 1	resid 1	fit 2	resid 2
Norway	250	95	109.75	−14.75	100.5	−5.5
Sweden	300	120	124.75	−4.75	120.5	−0.5
Denmark	350	165	139.75	25.25	140.5	24.5
Australia	470	170	175.75	−5.75	188.5	−18.5

x = cigarette consumption (per capita) y = lung cancer deaths (per million males)

Table 2. Data from "Smoking and Health", Report of the Advisory Committee to the Surgeon General of the Public Health Service, U.S. Department of Health, Education, and Welfare, Washington, D.C., Public Health Service Publication No. 1103, p. 176.

In Table 2, Columns 3 and 5 give the y values fitted by these two lines. Columns 4 and 6 give the differences between the actual y and the fitted y. For example, for the point (250, 95) and the first line, the fitted y is

$$109.75 = 34.75 + (.3)(250)$$

and the difference is $95 - 109.75$ or -14.75. The difference between an actual value and a fitted value is called a *residual.* Figure 4 shows the scatterplot with both lines.

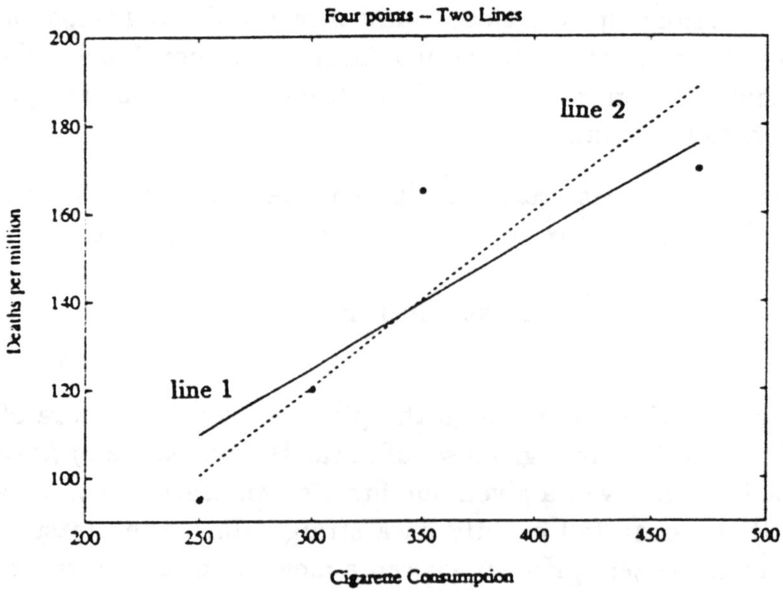

Figure 4.

Any line through the data provides us with four residuals. In assessing how well a particular line fits the data we would like these residuals (ignoring sign) to be collectively as small as possible. In general a line might make certain residuals very small (even zero if the line passes through that data point) but at the expense of increasing others. For example, line 1 is close to points 2 and 4 but farther from points 1 and 3, while line 2 is close to 1 and 2 but farther from 3 and 4. Overall, which line fits better?

We need a measure of fit that summarizes the four residuals into a single number. There are several ways we could define such a measure, but the most commonly used is called the *root mean square error* (RMSE). It is defined as the square root of the average of the squared residuals, e.g. for line 1 in our example,

$$\text{RMSE} = \sqrt{\frac{(-14.75)^2 + (-4.75)^2 + (25.25)^2 + (-5.75)^2}{4}} = 15.08.$$

For line 2, RMSE = 15.80. Based upon this measure of fit, line 1 fits the data better.

Our strategy now is to find that line that will minimize the RMSE. That is, of all possible lines through the scatterplot, which one has the smallest possible RMSE? This line is called the *least squares line* or the *regression line.*

Derivation of the Regression Line

We will find the line that minimizes the RMSE for a given data set. The first observation to make is that minimizing the RMSE is equivalent to minimizing the sum of the squared residuals. That is, if we have data $(x_1, y_1), (x_2, y_2), \ldots, (x_n, y_n)$, then we are searching for values a and b that minimize

$$\Big[y_1 - (a + bx_1)\Big]^2 + \Big[y_2 - (a + bx_2)\Big]^2 + \ldots + \Big[y_n - (a + bx_n)\Big]^2$$
$$= \sum_{i=1}^{n} \Big[y_i - (a + bx_i)\Big]^2.$$

Notice that in this minimization problem, the x_i's and y_i's are *constants*, the given data points. On the other hand, a and b are the *variables*, the quantities we are free to manipulate to minimize the sum of squared residuals.

One can approach this minimization problem in two ways. One could consider the quantity-to-be-minimized as a function of two variables denoted by, say, $G(a, b)$, and then use techniques for minimizing such functions. Instead we will restrict ourselves here to techniques of one-variable calculus by solving the problem in two stages.

First, we consider lines of a fixed slope, b. We want to find the value of a that minimizes

$$g(a) = \sum_{i=1}^{n} [y_i - (a + bx_i)]^2.$$

If you think about this as a function of a, remembering that all of the other letters represent fixed numbers, you'll see that it is a quadratic polynomial in a, in which the coefficient of a^2 is positive. Hence it has a unique critical point, which is a global minimum. To find the critical point, we calculate $g'(a)$:

$$\begin{aligned} g'(a) &= \frac{d}{da}\sum_{i=1}^{n}[y_i - a - bx_i]^2 = \sum_{i=1}^{n}\frac{d}{da}[y_i - a - bx_i]^2 \\ &= \sum_{i=1}^{n} 2[y_i - a - bx_i](-1) = -2\Big[\sum_{i=1}^{n} y_i - na - b\sum_{i=1}^{n} x_i\Big] \end{aligned}$$

If you are not used to working with the summation notation, the trickiest part of this calculation is to realize that $\sum_{i=1}^{n} a$ means we are adding up n terms, each of which is a, so the result is na. When we set $g'(a) = 0$ to find the critical point, the result is

$$a = \frac{1}{n}\sum_{i=1}^{n} y_i - b\frac{1}{n}\sum_{i=1}^{n} x_i.$$

Finally, recall that when we add n numbers and divide by n, we have found the *average* of those numbers. It is traditional to use the notation $\overline{y}$ for the average of the y_i's, and $\overline{x}$ for the average of the x_i's. Hence we can write our result as $a = \overline{y} - b\overline{x}$.

We have found that of all lines with a given slope b, the unique line which minimizes the RMSE is the line

$$y = (\overline{y} - b\overline{x}) + bx = \overline{y} + b(x - \overline{x}). \tag{1}$$

Notice that this is the unique line with slope b which passes through the point $(\overline{x}, \overline{y})$. The point $(\overline{x}, \overline{y})$ is sometimes called the *center of mass* of the scatterplot.

Now we know that we only need consider lines of the form (1). Among those lines, we'll find the value of b that minimizes the RMSE. We wish to find the minimum of

$$f(b) = \sum_{i=1}^{n}[y_i - (\overline{y} + b(x_i - \overline{x}))]^2.$$

Again, if you think about this as a function of b, remembering that all of the other symbols represent fixed numbers, you'll see that it is a quadratic polynomial in b, in which the coefficient of b^2 is positive. Hence it has a unique critical point, which is a global minimum. To find that point, we set the derivative equal to zero and solve for b:

$$f'(b) = \sum_{i=1}^{n} 2[y_i - (\overline{y} + b(x_i - \overline{x}))](-1)(x_i - \overline{x}) = 0$$

$$\sum_{i=1}^{n}(y_i - \overline{y})(x_i - \overline{x}) = b\sum_{i=1}^{n}(x_i - \overline{x})^2$$

$$b = \frac{\sum(x_i - \overline{x})(y_i - \overline{y})}{\sum(x_i - \overline{x})^2}. \tag{2}$$

This formula enables us to calculate the slope b for the regression line through any set of data points. We can then use $a = \overline{y} - b\overline{x}$ to find the y-intercept a, and the regression line is determined.

Here is an example of how the calculation works, using the data in Table 2:

	x_i	y_i	$x_i - \overline{x}$	$y_i - \overline{y}$	$(x_i - \overline{x})(y_i - \overline{y})$	$(x_i - \overline{x})^2$
Norway	250	95	−92.5	−42.5	3931.25	8556.3
Sweden	300	120	−42.5	−17.5	743.75	1806.3
Denmark	350	165	7.5	27.5	206.25	56.3
Australia	470	170	127.5	32.5	4143.75	16256.2
	$\overline{x} = 342.5$	$\overline{y} = 137.5$			sum = 9025	26675

$$b = \frac{9025.0}{26675} = .338 \qquad a = 137.5 - (.338)(342.5) = 21.6$$

Thus our regression line is given by

$$y = 21.6 + 338x.$$

Example. Let's find the regression lines for volume on diameter and volume on height for the cherry tree data. For diameter:

$$\overline{x} = 13.248, \quad \overline{y} = 30.171$$

$$\sum(x_i - \overline{x})(y_i - \overline{y}) = 1496.6$$

$$\sum(x_i - \overline{x})^2 = 295.44$$

$$b = \frac{1496.6}{295.44} = 5.07 \qquad a = 30.17 - (5.07)(13.25) = -36.9.$$

The regression line is $y = -36.9 + 5.07x$, with RMSE $= \sqrt{\dfrac{\sum(y_i - (a + bx_i))^2}{n}} = 4.252.$

Similarly, the regression line for volume on height is $y = -87.1 + 1.54x$ with RMSE $= 13.40$.

The scatterplots in Figures 1 and 2 indicated that volume was more strongly related to diameter than to height. The RMSE's of the respective lines quantify this difference in strength of relationship, since they estimate the average size of residuals from the lines. If we predict the volume of a tree using diameter, for a value of the diameter in the range we used to do the regression, we are likely to be off by an amount of about 4.25 cubic feet, while if we predict the volume using height we are likely to be off by about 13.40 cubic feet.

Exercises

1. Write out carefully and justify the intermediate steps in the derivation of formula (2).

2. Suppose a tree is found with diameter of 15 inches and height of 80 feet. Estimate its volume twice, using first the diameter and then the height. Which estimate is more accurate? Why?

3. Find the RMSE for the regression line for the smoking–lung cancer data, and verify that it is smaller than the RMSE for lines 1 and 2 in Figure 4.

*4. Using the data in the Appendix, find the regression line for estimating the delivery weights of mothers from their weights at conception. What would you estimate the delivery weight to be for a mother who weighs 142 pounds at conception? [Note: This estimate is chancy if the new mother is unlike the mothers in the sample in some fundamental way.]

5. On January 28, 1986, the twenty-fifth flight of the U.S. Space Shuttle program ended in disaster when one of the rocket boosters of the Shuttle *Challenger* exploded shortly after lift-off, killing all seven crew members. The Presidential Commission that investigated the accident concluded that it was caused by the failure of an O-ring in a field joint on the rocket booster, and that this failure was due to a faulty design that made the O-ring unacceptably sensitive to a number of factors, including temperature. The Commission learned that O-rings had been damaged on many of the 24 previous flights. The data on this damage are given in the Appendix. The missing data are from the fourth flight in which the hardware was lost at sea.

   (a) Plot 'Number of Damage Incidents' (y) against 'Temperature' (x). Describe any relationship that may exist between temperature and damage to O-rings. Would this relationship help you in deciding whether to trust the O-rings on a launch on a morning when the temperature was 31° F (the temperature on the morning of January 28, 1986)?

   (b) Unfortunately, when managers discussed the possible effects of cold weather, they only considered the data for the 7 flights for which the thermal distress of O-rings had occurred. Look at the scatterplot for these data and consider again the question in part (a).

   *(c) Even though a straight line model is only a crude approximation in this case, find the least squares lines in both (a) and (b). Estimate the number of incidents for 31 degree weather in each case. [Note: Extrapolation in a model beyond the range of x values upon which the model was derived is always risky business!]

Residual plots

When we fit a line $y = a + bx$ to our data, recall that the difference $e_i = y_i - (a + bx_i)$ between an actual y and a y predicted by the line, is called a ***residual.*** We think of the residuals $e_1, e_2, \ldots, e_n$ as estimates of error terms. If our model is adequate these residuals will appear random or patternless when plotted against the x_i values. Such a plot is called a ***residual plot.*** Figure 5a shows a scatterplot of delivery weight versus weight at conception for a sample of mothers. Figure 5b shows the residuals plotted against x. There is no particular pattern to this plot. This is characteristic of a residual plot where the straight line model is appropriate for the data.

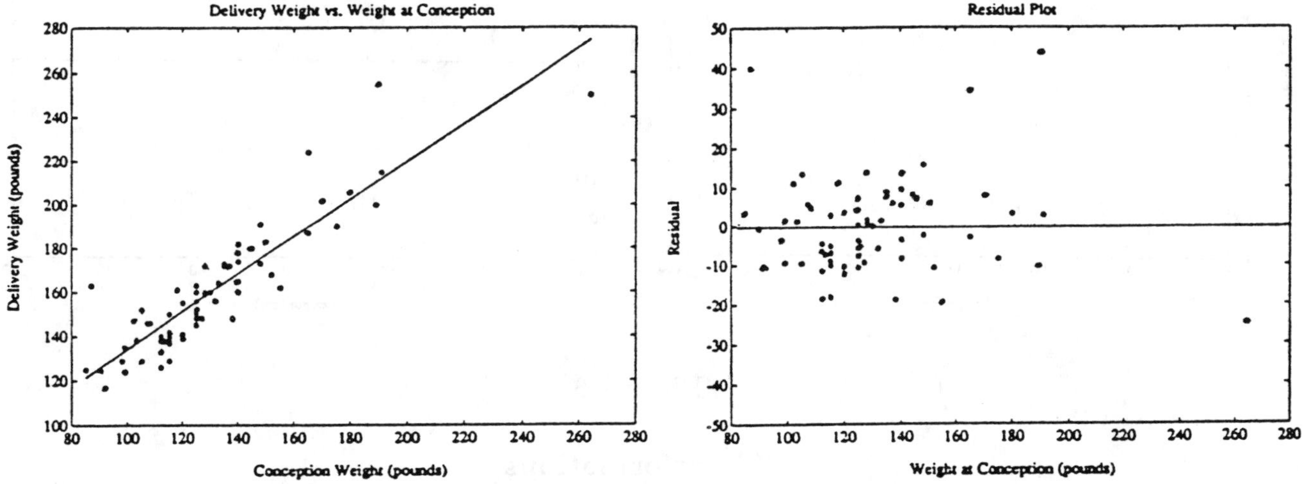

Figure 5.

Suppose, on the other hand, that there is a clear pattern in the residual plot. Figure 6a shows a scatterplot for experimentally determined measurements on the period of a swinging pendulum compared to its length. A regression is performed. Clearly the fit is very good: the RMSE is only 0.025. However, the residual plot in Figure 6b shows a clear pattern of increasing and then decreasing residuals, indicating a downward concavity in the scatterplot. Yet this concavity is not so visible in the scatterplot. The residual plot accentuates the discrepancies between data and a model. This pattern in the residuals tells us the straight line model doesn't fully capture the relationship between y and x.

Of course, physics tells us that we shouldn't expect a straight line to fit the pendulum data. A better model (still approximate) derived from physical theory is that

$$y = \frac{2\pi}{\sqrt{g}}\sqrt{x},$$

where y = period (seconds), x = length (cm) and g = gravitational acceleration ≈ 980 cm/sec^2. Of course, if we knew the theory, we wouldn't be using regression to predict period from length. But we could be using the regression either to check the theory or to estimate the gravitational constant g. We will discuss fitting such a non-linear model in the next section.

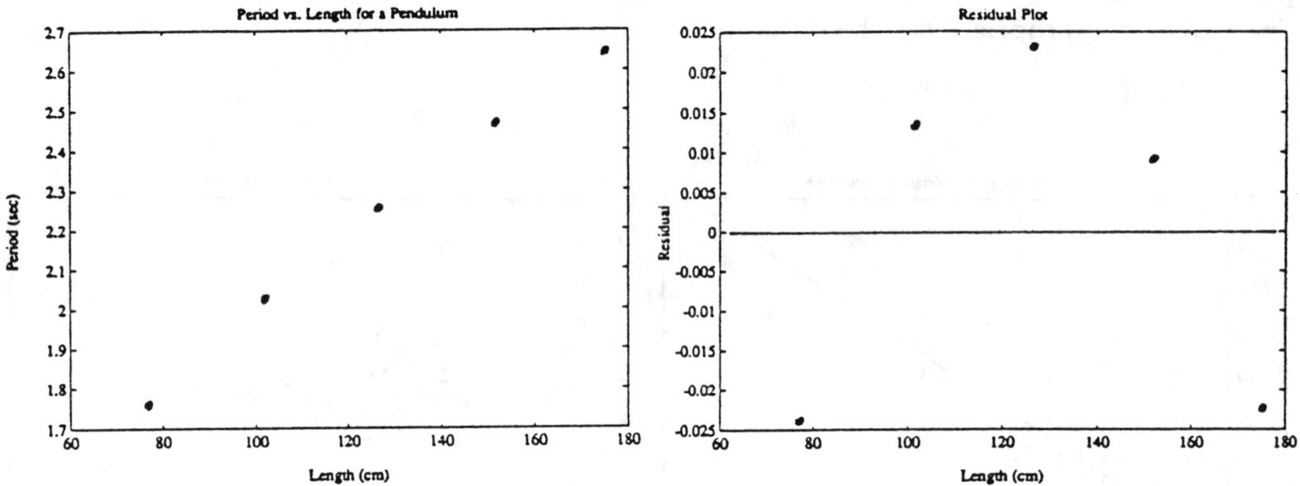

Figure 6.

Transformations

Sometimes theory will tell you that y is not linearly dependent upon x. Sometimes you will reach this conclusion, not from theory, but empirically by using a residual plot. In either case we need methods for fitting more complicated models to data. One such method involves transforming either y or x to *linearize* the data. Transforming a variable simply means taking a function of it. We will consider some representative examples.

From the data on the number of mammary cancer cells in an experimental culture in Figure 3, we suspected that the underlying relationship might be of the form

$$y = ae^{bx}$$

for some unknown a and b. If so, then it will be true that

$$\log y = \log a + bx = a' + bx.$$

This suggests that we transform y via $y' = \log y$. (We consistently use natural logarithms in this module. Also, y' is the traditional name for the new variable introduced by transforming y—the $'$ here does not indicate a derivative.) Figure 7 shows the log of population versus time and its residual plot. Notice that the log transformation has linearized the data. Both the scatterplot and the residual plot show this.

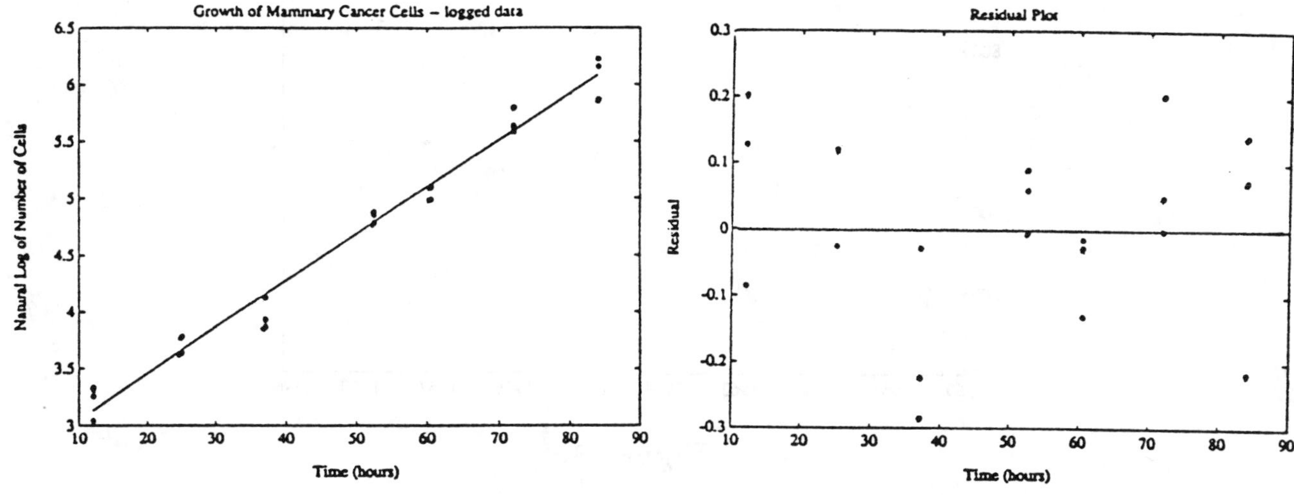

Figure 7.

To find a good-fitting exponential curve $y = ae^{bx}$ to the data in Figure 3, we find a and b by fitting a straight line $y' = a' + bx$ to the data in Figure 7 and then take

$$a = e^{a'}.$$

The best fit line for Figure 7 turns out to be $\log y = 2.64 + .0410x$. Hence we have $a = e^{2.64} = 14.0$, so that the best fit exponential function becomes

$$y = 14e^{.0410x}.$$

You are asked to do another simple transform, this time of x, in Exercise 6. For an example of a more complicated transform, consider the growth of the American intercontinental ballistic missile force during the 1960's (Figure 8). This data has the S shape characteristic of the logistic model for constrained growth of a population.

The logistic curve is $y = L/(1 + e^{a+bx})$, where $y =$ population at time x, and L, a and b are constants. L is called the *carrying capacity* of the environment. Note that

$$\frac{L}{y} = 1 + e^{a+bx}$$

$$\frac{L}{y} - 1 = \frac{L - y}{y} = e^{a+bx}$$

$$\log\left(\frac{L - y}{y}\right) = a + bx.$$

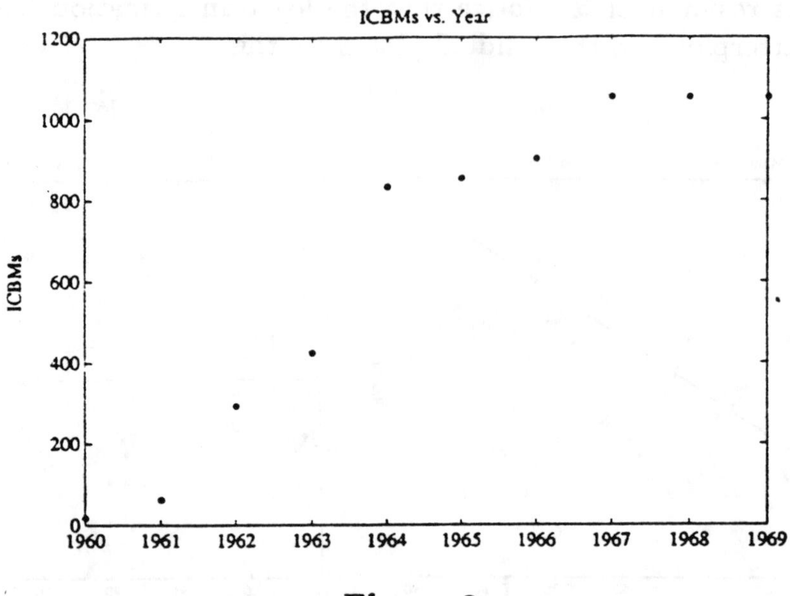

Figure 8.

Hence, if one can estimate L, the transformation $y' = \log\left(\frac{L-y}{y}\right)$ linearizes the relationship. For this data, the estimate of $L = 1060$ seems to provide a good fit. Figure 9 gives the transformed data and the residuals from regression on the transformed data, showing the good fit of the logistic model. The final estimated model for this growth data is

$$y = 1060/(1 + e^{2092-1.0654x}).$$

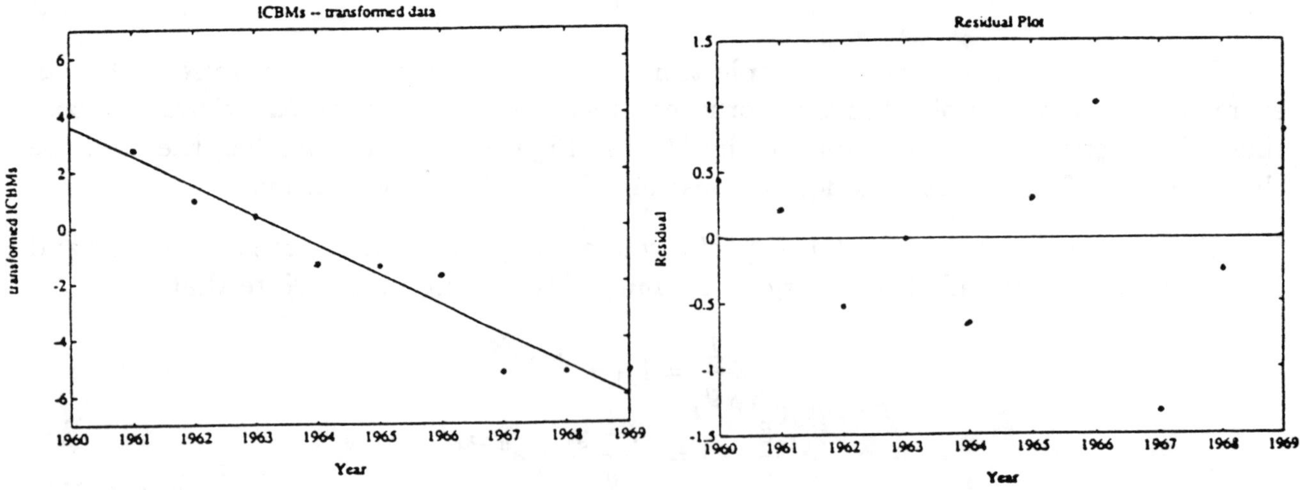

Figure 9.

Exercises

6. Transform the pendulum data (in the Appendix) to linearity by considering the transform that physical theory suggests: $x' = \sqrt{x}$. Compute the least squares line on the transformed data: $y = a + bx'$. Plot the residuals. Do they appear reasonably random? From the data, what is your estimate of the gravitational constant?

*7. Use $L = 1100$ in the logistic transform for the ICBM's in Figure 10 (data in the appendix). Find the regression line for the tranformed data. Plot the residuals against time. Does this model fit as well as with $L = 1060$?

References

[GW] Goldman, Robert N. and Weinberg, Joel S. (1985), *Statistics: An Introduction*, Prentice-Hall.

[LM] Larsen, Richard J. and Marx, Morris L. (1985), *An Introduction to Mathematical Statistics and Its Applications*, Second Edition, Prentice-Hall.

[RJR] Ryan, Barbara F., Joiner, Brian L. and Ryan, Thomas A., (1985), *Minitab*, Second Edition, Duxbury Press.

Data Appendix

Pregnancy Weight Data

Weight at Conception (pounds)	Weight at Delivery (pounds)	Weight at Conception (pounds)	Weight at Delivery (pounds)
115	140	120	155
112	126	140	182
125	145	133	164
108	146	92	117
112	133	165	224
115	137	125	148
99	135	118	161
140	178	135	173
115	150	127	148
135	172	87	163
103	138	165	187
191	215	112	138

Source of data: [GW], p. 48.

ICBM Data

Year, x	Number of ICBM's, y
1960	18
1961	63
1962	294
1963	424
1964	834
1965	854
1966	904
1967	1054
1968	1054
1969	1054

Source: [LM]

Challenger Data

Flight	Date	Temperature	Number of Damage Incidents
STS–1	4/12/81	66	0
STS–2	11/12/81	70	1
STS–3	3/22/82	69	0
STS–4	6/27/82	80	(data not available)
STS–5	1/11/82	68	0
STS–6	4/4/83	67	0
STS–7	6/18/83	72	0
STS–8	8/30/83	73	0
STS–9	11/28/83	70	0
STS 41–B	2/3/84	57	1
STS 41–C	4/6/84	63	1
STS 41–D	8/30/84	70	1
STS 41–G	10/5/84	78	0
STS 51–A	11/8/84	67	0
STS 51–C	1/24/85	53	3
STS 51–D	4/12/85	67	0
STS 51–B	4/29/85	75	0
STS 51–G	6/17/85	70	0
STS 51–F	7/29/85	81	0
STS 51–I	8/27/85	76	0
STS 51–J	10/3/85	79	0
STS 61–A	10/30/85	75	2
STS 61–B	11/26/85	76	0
STS 61–C	1/12/86	58	1
STS 51–L	1/28/86	31	(Challenger accident)

Source: "The Report of the Presidential Commission on the Space Shuttle Challenger Accident," 1986.

Pendulum Data

Length	Period
175.2	2.650
151.5	2.468
126.4	2.256
101.7	2.024
77.0	1.764

Length = length of pendulum in centimeters.

Period = period of the pendulum based upon the average of 50 cycles (in seconds).

Source: [RJR], p. 253.

Answers to Exercises

2. An estimate from the diameter is 39 ± 4 cubic feet; from the height, 36 ± 13 cubic feet.

3. RMSE $= 14.76$.

4. The regression line is $y = 40 + .944x$, where $x =$ weight at conception and $y =$ weight at delivery. When $x = 142, y = 174$.

5. The data is plotted below. There is a clear upward trend to the left: the O-rings are more likely to fail at lower temperatures. Ignoring the points at height zero almost completely veils this trend. With all the points, the regression line is $y = 4.8 - .0627x$, which predicts $y = 2.85$ when $x = 31$. Without the points at height zero, the regression line is $y = 3.05 - .0254x$, showing the much weaker dependence on temperature. When $x = 31$, $y = 2.26$, not significantly higher at the lower temperature.

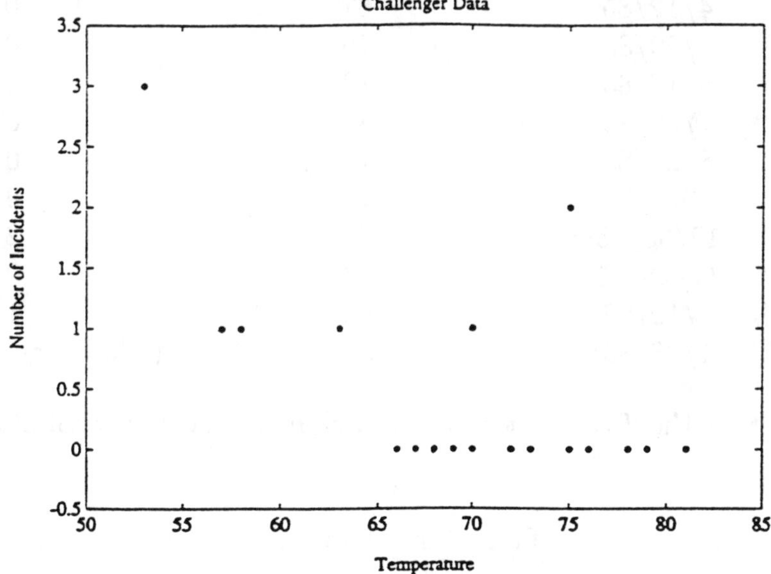

6. The regression line for the transformed data is $y = .02 + .199x' = .02 + .199\sqrt{x}$. The residuals do look fairly random (although it is always possible to see some pattern in just 5 points). To estimate g,

$$\frac{2\pi}{\sqrt{g}} = b = .199, \text{ so } g = \left(\frac{2\pi}{.199}\right)^2 = 997 \text{ cm/sec}^2,$$

which compares fairly well to the value $g = 980$ cm/sec^2.

7. The regression line after transformation is $y' = 1587 - .808x$, so the fitted logistic equation is

$$y = \frac{1100}{1 + e^{1587 - .808x}}.$$

However, the residual plot for the transformed data, given below, shows a definite concave-up pattern, indicating that $L = 1100$ is not a good choice for L.

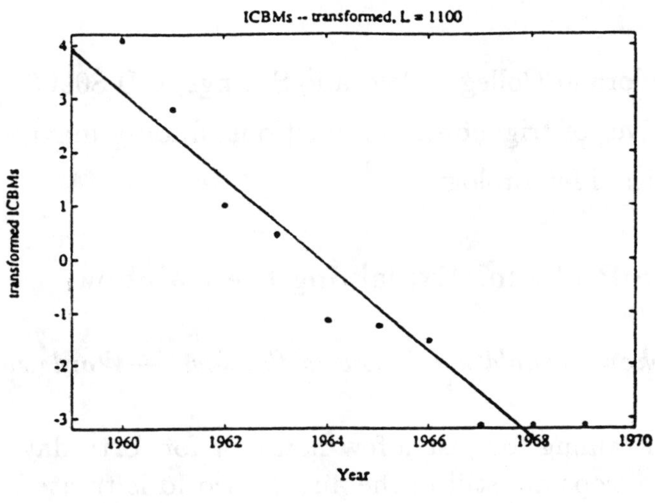

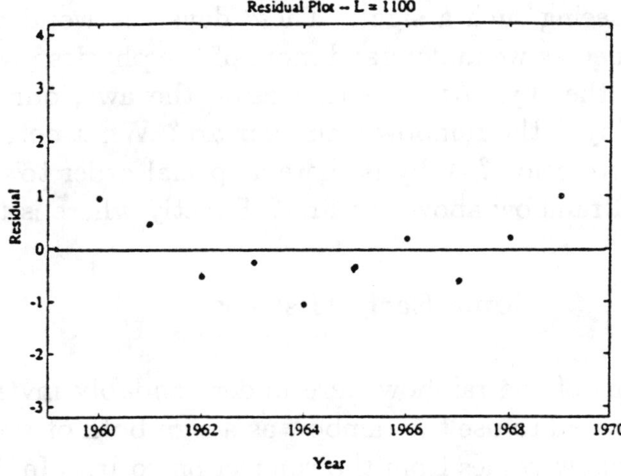

SOMEWHERE WITHIN THE RAINBOW

Author: Steven Janke, Colorado College, Colorado Springs, CO 80903

Calculus Needed: Derivatives of trigonometric functions, finding maxima and minima.

Area of Application: Optics, Meteorology.

The Problem: Explaining the Rainbow

My heart leaps up when I behold a rainbow in the sky...— Wordsworth

Whether it has been raining for just a few hours or for forty days and forty nights, if the sun appears and raindrops are still in the air, the world is treated to one of nature's most vivid spectacles, the rainbow. Imagine the mixture of fear and wonder that ancient people must have felt on seeing such a sight. These days the wonder is still there, but certainly the fear has lessened as we understand more of the physics involved in producing such a display of color in the sky. After setting aside the awe, our curiosity produces question after question. Why is the rainbow a circular arc? What determines how high it is in the sky? Why are there colors? Why is there a special order to the colors? Why is there occasionally a second rainbow above the first? Exactly where is the pot of gold?

Some Early History

The early explanations of the rainbow were understandably mythological in origin. The Greek goddess Iris was said to use the rainbow as a sign both of warning and of hope. The word "iridescent" probably comes from the connection to Iris. In African mythology, the rainbow was a large snake coming out to graze after the storm. Here again the event is both a sign of hope and one of fear, for the snake could gobble children that were too close to the ends of the bow. The ends do appear to touch the earth leading some to claim that great treasure was buried there. Yet in a less capitalistic vein, many American Indians saw the bow as a bridge anchored in this world and leading to the next.

In 578 B.C., Anaximenes, a Greek scholar, noted the relation between the rainbow and the sun. Rather than attributing the bow to celestial powers, he suggested that clouds bent the sun's light to produce the arc of colors. Aristotle used careful geometry, but faulty reflection laws, to establish the circular shape of the bow. Gradually, scholars began to see that both reflection and refraction of light had something to do with the rainbow phenomenon. In the fourteenth century, Theodoric of Frieburg and the Persian scholar Kamal al-Din al Farisi independently decided that drops of rain were the key. They looked closely at the way a globe of water affected light and were able to give correct qualitative explanations for the bow.

The rainbow has piqued the interest of many scholars in each of the last several centuries. The sixteenth century seems to have produced the most books on the subject, but few of them were of major importance. As you might expect, seventeenth century scholars like Kepler, Descartes, Fermat, and Newton all made significant contributions to the study of the rainbow. Even today, physicists continue to tidy up the theory. Understanding the rainbow is so tied with understanding the nature of light that until theories of light are complete, there will be open questions about the rainbow.

Reflection

Light from the sun, refracted and reflected by water droplets in the atmosphere, forms the rainbow, so the first step in explaining the phenomenon is to understand how light is bent by various substances. In 1657, the extraordinary mathematician Pierre de Fermat turned his attention to the bending of light and proved the main results by postulating a simple principle. Fermat suggested that in traveling from point P to Q, light follows a path which minimizes the total travel time.

Fermat's Principle. *Light follows a path which minimizes the total travel time.*

Consider first the reflection of light. It helps when discussing geometric problems with light to imagine that light travels along rays. So suppose we have a source of light rays at point P in Figure 1. Imagine that we detect one of the rays passing through point Q after reflecting off a surface. At what point R does the ray reflect off the surface?

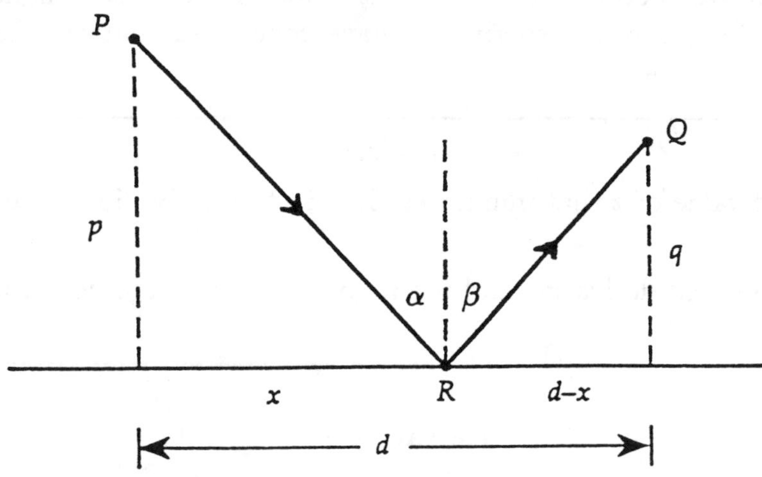

Figure 1.

Fermat's principle claims the ray follows a path that minimizes the time necessary to travel from P to Q while reflecting off the surface. Assuming the speed of light in our example is constant, the point R should be positioned so the path PRQ has minimum length. Considering the triangles in the figure, we get the following expression for the path length as a function of x:

$$L(x) = \sqrt{p^2 + x^2} + \sqrt{q^2 + (d - x)^2}.$$

To find the minimum path length, we find the derivative $L'(x)$ and set it equal to zero.

$$\begin{aligned} L'(x) &= \frac{1}{2}(p^2 + x^2)^{-\frac{1}{2}}(2x) + \frac{1}{2}(q^2 + (d-x)^2)^{-\frac{1}{2}} \cdot 2(d-x) \cdot (-1) \\ &= \frac{x}{\sqrt{p^2 + x^2}} - \frac{(d-x)}{\sqrt{q^2 + (d-x)^2}} = 0, \\ \text{so that } \quad \frac{x}{\sqrt{p^2 + x^2}} &= \frac{(d-x)}{\sqrt{q^2 + (d-x)^2}}. \end{aligned}$$

Now referring again to the figure, $\sin\alpha = \dfrac{x}{\sqrt{p^2 + x^2}}$. Similarly, $\sin\beta = \dfrac{d-x}{\sqrt{q^2 + (d-x)^2}}$. So $L'(x) = 0$ when $\sin\alpha = \sin\beta$. We should verify that this is a minimum by taking the second derivative (see the exercises). Rather than actually solve for the distance x, it is more useful to note that the minimum occurs when the sines of the two angles are the same. Since the angles are both between 0 and $\pi/2$, we conclude that the two angles are equal. For convenience call α the angle of incidence and β the angle of reflection.

Law of Reflection. *For reflection, the angle of incidence is equal to the angle of reflection.*

Note that we have deduced the Law of Reflection from Fermat's principle of least time. We have not proved Fermat's principle, but it does make sense in light of other results in physics. And in fact, careful experiments have concluded that the Law of Reflection does hold.

Exercises

1. Determine the value of x that minimizes $L(x)$ in the derivation of the Law of Reflection.
2. Compute the second derivative and use it to show that we indeed found a minimum for $L(x)$.

Refraction

When dealing with reflection, we assumed that the light rays were traveling only in air and therefore maintained a constant speed. However, to attack the rainbow questions, we need to also understand what happens when light travels through water. It turns out that the speed of light in water is less than the speed in air. Our derivation of the reflection law would be identical for a mirror and light source submerged in water since the speed of light would again be constant, but what happens if part of the light's path is in water and part is in air?

Figure 2 shows a new setup where point P is again a source of light rays in air. Now, however, point Q is in water. We are interested in the path of a light ray that leaves P and passes through Q. It crosses the air/water interface at the point R. The angle the path PR makes with the line perpendicular to the water's surface is called the angle of incidence and is represented by α. The corresponding angle between the path RQ and the perpendicular is called the angle of refraction and is represented by β. Fermat's principle claims that the point R is positioned so as to make the total time of travel a minimum. Since the speed changes when the light crosses into water, we need to consider both speeds in our analysis.

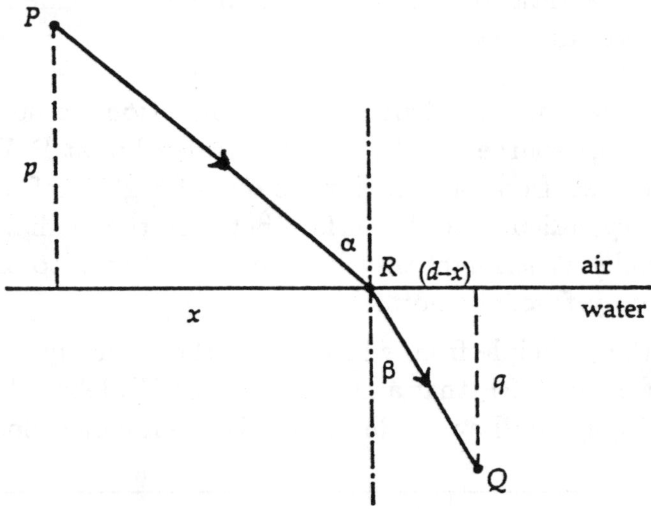

Figure 2.

Let c_a be the speed of light in air and let c_w be the speed of light in water. Remembering that time is distance divided by speed, we calculate that the light ray spends $\dfrac{\sqrt{p^2+x^2}}{c_a}$ units of time traveling from P to R and $\dfrac{\sqrt{q^2+(d-x)^2}}{c_w}$ units of time traveling from R to Q. Thus the total time is

$$T(x) = \frac{\sqrt{p^2+x^2}}{c_a} + \frac{\sqrt{q^2+(d-x)^2}}{c_w}.$$

Again to find the minimum, we take the derivative of $T(x)$:

$$\begin{aligned} T'(x) &= \frac{1}{c_a} \cdot \frac{x}{\sqrt{p^2+x^2}} - \frac{1}{c_w} \cdot \frac{d-x}{\sqrt{q^2+(d-x)^2}} \\ &= \frac{\sin\alpha}{c_a} - \frac{\sin\beta}{c_w}. \end{aligned}$$

Setting $T'(x) = 0$ gives

$$\frac{\sin\alpha}{c_a} = \frac{\sin\beta}{c_w}, \quad \text{or} \quad \frac{\sin\alpha}{\sin\beta} = \frac{c_a}{c_w}.$$

In other words the ratio of the sines is a constant. Note again that in order to verify that we have found the minimum we should take the second derivative.

This constant c_a/c_w is the ratio of the speed of light in air to the speed of light in water. In order to calculate it, tables have been compiled that give the ratio of the speed of light in a vacuum to the speed of light in various media. For example, the ratio of the speed in a vacuum to the speed in water is about 1.33 and is called the ***index of refraction*** for water. The index of refraction for air is very close to 1 so the ratio c_a/c_w is very close to 1.33.

Law of Refraction. *The ratio of the sine of the angle of incidence to the sine of the angle of refraction is a constant.*

In our derivation, there was no dependence on direction, so our result would be the same if we assumed that the source of light was at Q instead of at P. With this observation we notice that if light travels from one medium to one of higher refractive index, the light ray bends toward the perpendicular to the surface between the media. (This perpendicular is often called the normal.) When light travels from one medium to one of lower refractive index, the ray is bent away from the normal.

Fermat supplied the principle from which we mathematically deduce the Law of Refraction, but it was earlier, in 1621, that a Dutch scientist Willebrord Snell experimentally discovered the result. Today the Law of Refraction is often called Snell's law.

Exercise

3. Verify that we found a minimum for $T(x)$ in the derivation of the Law of Refraction.

The Rainbow Angle

Rainbows form when raindrops both reflect and refract light from the sun. When light traveling through the air strikes a drop, some of the light is reflected and some is refracted as it enters the drop. Part of the light inside the drop is reflected when it strikes the other side of the drop and part is refracted as it again passes into the air. In general, when light travels from one medium to another, part of the light is reflected at the interface and part continues into the second medium where it is refracted. To understand how the rainbow forms, we need to keep track of the reflections and refractions caused by a drop of rain.

The shape of a raindrop depends on several factors, but for a good approximation, it is fairly safe to assume that it is spherical. Look then at Figure 3. Here we see the cross-section of a drop as a light ray enters it at point A. Some of the light ray will be reflected, but the figure shows the part that enters the drop. The Law of Refraction says that this ray will be bent toward the normal since the refractive index of water is larger than that of air. From geometry, we know that the tangent to the circle at point A is perpendicular to the radius of the circle through A. Hence, the radius through A is the normal at A. In the figure, α is the angle of incidence and β is the angle of refraction.

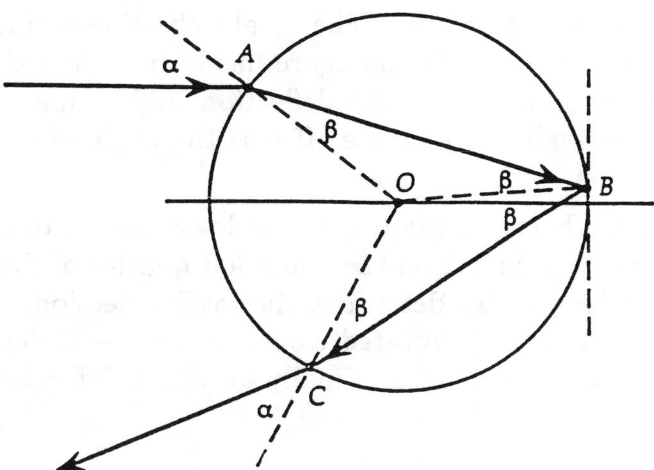

Figure 3.

The ray continues through the drop and strikes the other side at point B. Here again, part of the ray is reflected and part continues into the air where it is refracted. In the figure, we follow the reflected part. At B, the ray is reflected so that the angle of incidence equals the angle of reflection. Here the angle of incidence is the angle between the ray and the tangent at B. Notice that this implies that angle ABO equals angle OBC. When the ray hits the drop's surface at C, part is reflected, but let's follow the part that enters the air and is refracted. Since the ray moves into a medium of lower refractive index, it is bent away from the normal.

Figure 3 traces parts of one particular ray that strikes the drop. At each interface, another part is either refracted or reflected, and consequently there are many paths a ray could take in interacting with the drop. In fact, you can imagine a ray that enters the drop and is repeatedly reflected around inside it. Since at each interface between air and water part of the ray is reflected and part refracted, when we choose to follow one part we are following a ray that has less intensity than the original ray. Each time an interface is hit, the light intensity decreases. We are therefore interested in rays that strike the interface only a few times, for this will be the brightest light.

Again looking at the figure, a ray that strikes the drop at A and is simply reflected will be fairly bright, but as we will see, such a ray does not add to the essential features of the rainbow since it doesn't interact with the water. Similarly, a ray that hits at A and then travels through the drop to exit at B will also be fairly bright, but we would have to be on the righthand side of the drop to see this light. Rainbows are formed when the sun is behind us and light from it is reflected in various ways from the raindrops. So the ray drawn in the figure is the simplest ray involved in rainbow formation.

Notice that the point A could be anywhere on the left half of the circle. If it is on the upper half of the circle then the ray exits the drop in the lower half. We are interested in how much the ray is deflected once it exits the drop. For example, if the ray comes in the drop along the diameter of the circle, then the angle of incidence is zero and therefore the angle of refraction is zero. The ray will reflect off the back of the drop and exit the drop along the same diameter that it entered on. The total deflection would be 180 degrees in a clockwise direction. The ray drawn in the figure has been deflected by less than 180 degrees. As the point A moves on the circle, the deflection angle changes. So the angle of deflection is a function of the angle of incidence. If α is the angle of incidence, let $D(\alpha)$ represent the angle of deflection.

Because of the symmetry between the upper and lower halves of the the circle, we might as well focus only on those points A on the upper-left quarter of the circle. For these points, α varies from 0 to 90 degrees. To determine the total deflection, first consider how the ray is deflected at the point A. It is rotated clockwise by $\alpha - \beta$ degrees. At B, it is again rotated clockwise by $180 - 2\beta$ degrees. Finally at C the deflection is again $\alpha - \beta$ degrees. Hence

$$D(\alpha) = \alpha - \beta + 180 - 2\beta + \alpha - \beta = 180 + 2\alpha - 4\beta.$$

Notice that D is a function of both α and β. However, we know from the Law of Refraction that β can be expressed as a function of α. We will need to keep this in mind when we take the derivative.

Now $D(C) = 180$ and as α increases, $D(\alpha)$ at first decreases. But what is interesting is that D has a minimum. It only decreases so far and then it increases. To find this minimum, we take the derivative (recalling the chain rule) and get

$$D'(\alpha) = 2 - 4\frac{d\beta}{d\alpha}.$$

Remember that, from the Law of Refraction,

$$\frac{\sin\alpha}{\sin\beta} = k \text{ where } k = \frac{c_a}{c_w}.$$

If we differentiate this with respect to α we get

$$\cos\alpha = k\cos\beta \cdot \frac{d\beta}{d\alpha}.$$

Solving for $d\beta/d\alpha$ and substituting into the expression for $D'(\alpha)$ gives

$$D'(\alpha) = 2 - 4\frac{\cos\alpha}{k\cos\beta}.$$

Setting the derivative equal to zero we have,

$$D'(\alpha) = 2 - \frac{4}{k} \cdot \frac{\cos\alpha}{\cos\beta} = 0$$

$$\text{which implies} \quad \frac{k}{2} = \frac{\cos\alpha}{\cos\beta}.$$

We want the value of α which satisfies this equation, so we eliminate β. Squaring both sides gives

$$\frac{k^2}{4} = \frac{\cos^2\alpha}{\cos^2\beta} = \frac{\cos^2\alpha}{1-\sin^2\beta}.$$

But $\sin\beta = \frac{1}{k}\sin\alpha$, so

$$\begin{aligned} \frac{k^2}{4} &= \frac{\cos^2\alpha}{1-\frac{\sin^2\alpha}{k^2}} \\ \frac{1}{4} &= \frac{\cos^2\alpha}{k^2-\sin^2\alpha} = \frac{\cos^2\alpha}{k^2-(1-\cos^2\alpha)} \\ k^2-1+\cos^2\alpha &= 4\cos^2\alpha \\ \cos\alpha &= \sqrt{\frac{k^2-1}{3}}. \end{aligned}$$

Finally we have an expression for the cosine of the incidence angle with minimum deflection. Since raindrops are water, $k \approx 1.33$, so $\cos\alpha \approx 0.5063$ and $\alpha \approx 59.56°$. At this incidence angle, the deflection is $D(59.56) \approx 137.5°$. To establish that this is a minimum we can check the sign of the second derivative (see the exercises).

We have found the incidence angle, $\alpha \approx 59.58°$, that gives the minimum deflection. Since the derivative of the deflection function is zero at this special angle, we know that the change in deflection angle divided by the change in incidence angle is nearly zero near $\alpha \approx 59.58°$. In other words, many rays with incidence angle near 59.58° get deflected by about the same amount. Rays further away from this critical angle get spread out more. So if we are looking at the deflected light, then rays coming from the direction of minimum deflection should appear the brightest since they are spread out the least. This is where the rainbow appears. The ray whose incidence angle is $\alpha \approx 59.58°$ is called the *rainbow ray* and 42.5° (= 180 − 137.5) is called the *rainbow angle*. The rainbow angle is the angle from the horizontal at which an observer should see the rainbow, if the rays of sunlight are horizontal. Figure 4 shows how the rainbow angle is related to the sun, observer, and the raindrops.

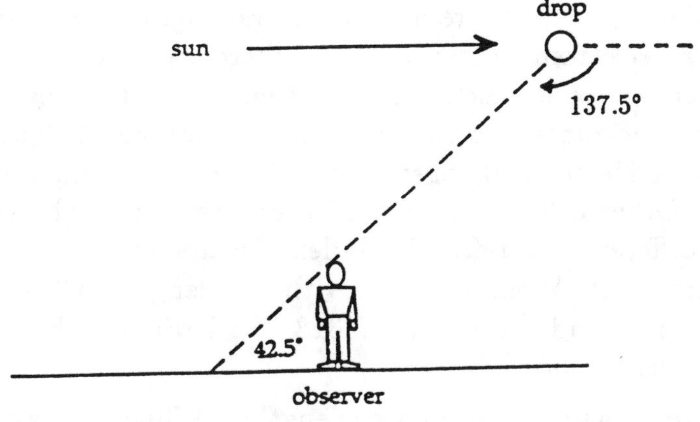

Figure 4.

Drops inclined 42.5° from an observer appear brighter than those with less inclination. For drops higher in the sky, the deflection angle would have to be less than 137.5° and since we discovered this angle is a minimum, no rays of the type we have been tracing come from drops higher in the sky. Any light coming from high in the sky must come from rays that have more than one (or none) internal reflections.

In the early part of the seventeenth century, Descartes carried out an analysis leading to discovery of the rainbow angle. Since the techniques of calculus were not available to him, he had to calculate the deflection of many different rays and even then did not have a nice expression for the incidence angle of minimum deflection.

Now that we know that light scattered by a drop is brighter at a certain angle of observation, any drop in the sky at the correct angle will show some brightness. Imagine the observer at the vertex of a cone with vertex angle equal to twice the rainbow angle. Cutting the cone with a plane perpendicular to its axis gives a circular cross-section and every raindrop on this circle forms the rainbow angle with the observer. Consequently, the observer should see a bright circular arc in the sky. This is the rainbow. Notice that the rainbow may be higher or lower in the sky depending on how high the sun is. To an observer on the ground, the rainbow is at most one half of a circle. However, to an observer flying in a plane, the rainbow may form an entire circle.

Exercises

4. Verify that we found a minimum deflection angle by checking the second derivative. (Hint: You can find the second derivative by first finding the second derivative of β with respect to α. The trigonometric formula for $\sin(\beta - \alpha)$ will be helpful.)
5. Sketch the function $D(\alpha)$ for α between 0 and 90 degrees.
6. If an observer sees the rainbow at an angle of 25 degrees from the horizontal, what is the sun's angle of inclination?

Colors

The geometry of light rays that we have considered so far accounts for a circular arc of brighter light in the sky. But where are the colors? Actually the answer is quite simple now. Light is really an electromagnetic wave and therefore we can talk about its frequency and wavelength. There is a wide spectrum of wavelengths, but our eyes are sensitive only to wavelengths in the range from about 7000 angstroms to about 4240 angstroms. Light with a wavelength of about 6470 to 7000 angstroms is perceived as red, and light in the 4000 to 4240 range is violet. Other colors fall between these two. Since the wave characteristic of these two colors are different, the refractive index of water varies depending on which color of light is passing through it. When red light with wavelength 6563 angstroms travels from air to water, the refractive index is about 1.3318. With violet light (4047 angstroms), the index increases to about 1.3435.

Sunlight is really a wide range of wavelengths. When it strikes a raindrop, wavelengths in the red range are refracted differently from those in the violet range. The other colors like blue and yellow fall between these two ranges and are refracted to various degrees between the two extremes. Consequently the light is actually spread into its constituent colors.

Now we need to repeat the calculation done to find the minimum angle of deflection. For red light, the minimum deflection is 137.7° and for violet light it is 139.4°. These values give rainbow angles of 42.3° and 40.6° respectively. In other words, when looking in the sky, the observer will see a circular arc of red light at a slightly higher inclination than the circular arc of violet light. The other wavelengths that we recognize as colors will form bows between these two. The order is red, orange, yellow, green, blue, indigo, and violet. (Taking the first letters gives a mnemonic: ROY G. BIV).

Newton was the first to make these careful calculations that explain the colors in the rainbow. By subtracting the rainbow angles for red and violet light it looks like the width of the bow is 1.7 degrees. Actually all these results assume that the rays from the sun are all parallel. To correct for the fact that the rays are not quite parallel, Newton allowed 0.5 degrees for the angular diameter of the sun and concluded that the rainbow width should be 2.2 degrees. This is in good agreement with actual observation although as we shall see later, the width of the bow does vary.

The Secondary Bow

Recall that the rainbow ray we traced was reflected once by the back of the raindrop. Other rays are reflected several times inside the drop. Each reflection reduces the intensity of the ray, but it is worth tracking at least those rays that have two internal reflections. To do this, look at Figure 5.

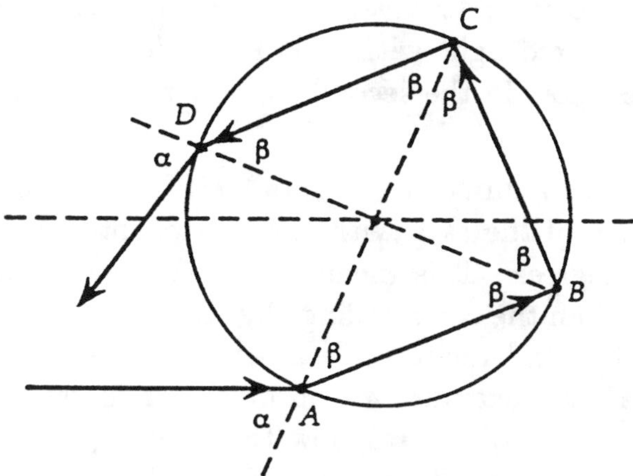

Figure 5.

This time we will follow rays incident on the bottom half of the drop since these rays are the ones that reach the observer. Keeping track of the deflections, we notice that the ray is rotated counter-clockwise at each of the points A, B, C, and D. The amount of rotation is similar to the analysis we did before, so this time we get

$$\begin{aligned}\text{Total deflection} &= (\alpha - \beta) + (180 - 2\beta) + (180 - 2\beta) + (\alpha - \beta)\\ &= 360 + 2\alpha - 6\beta.\end{aligned}$$

Since a 360 degree deflection means the ray continues in the same direction it started in, we can disregard the 360 and consider the deflection to be $2\alpha - 6\beta$. However, this is a counter-clockwise deflection and in order to compare it to the deflection for the rays with single internal reflections, we need to change this to a clockwise deflection. This is easily done by multiplying by -1. This gives us a new deflection function, D_2, for rays with two internal reflections:

$$D_2(\alpha) = 6\beta - 2\alpha.$$

Notice that $D_2(0) = 0$ and that D_2 begins to increase as α increases. In order to determine if this trend continues, we find any critical points by taking the derivative and setting it equal to zero. This time the critical point satisfies

$$\cos\alpha = \sqrt{\frac{k^2 - 1}{8}}.$$

With $k = 1.33$, we obtain the critical point $\alpha = 71.94°$, and $D_2(71.94) = 129.9°$. At this new critical point, D_2 is actually a maximum.

Hence for rays with two internal reflections, the maximum deflection angle is about 130°. In other words, raindrops that are inclined about 50°(i.e. 180° − 130°) from the observer will appear bright, although not as bright as those at 42°. This secondary arc of brightness is another bow which is dimmer that the primary bow and, unless conditions are right, is often too dim to see. Moreover, since D_2 is concave down, when we compare the maximum deflection for red light with that for violet light, we find that red light is deflected the most so the colors in the secondary bow appear in reverse order from those in the primary bow.

Notice also that the maximum of D_2 is about 130° while the minimum of D is about 138°. In other words, none of the rays with one or two internal reflections are deflected in the range 130 to 138 degrees. This means that the region between the primary and secondary bows is darker than the surrounding sky. It isn't totally black since light comes from rays that are reflected and refracted in many other ways. This darkened band is called Alexander's band after Alexander of Aphrodisias, a follower of Aristotle. Alexander deduced from Aristotle's theory of the rainbow that the region between the bows should be particularly bright. Since it wasn't, Alexander saw the need for a revised theory even though he couldn't supply one.

Exercises

7. Verify that the critical point for D_2 does occur at the point where $\cos\alpha = \sqrt{\frac{k^2 - 1}{8}}$.

8. Sketch the graph of D_2.

9. Determine the maximum deflection angle for red light and violet light.
10. Using the same procedure as above, find the deflection function D_n for rays that have n internal reflections. Find the critical point for this function. Theoretically, each of these classes of rays gives rise to another rainbow. They are rarely seen in the sky because they are so dim, but often one can see the first few bows in a laboratory set-up.

References

Boyer, Carl (1959), *The Rainbow from Myth to Mathematics*, Thomas Yoseloff Publishing, New York.

Greenler, Robert (1980), *Rainbows, Halos, and Glories*, Cambridge University Press, Cambridge.

Meyer-Arendt, Jurgen (1972), *Introduction to Classical and Modern Optics*, Prentice-Hall, New Jersey.

Nussenzveig, H. (1977), "The Theory of the Rainbow," *Scientific American*, April.

Tricker, R. (1970), *Introduction to Meteorological Optics*, American Elsevier, New York.

A longer version of this module appeared in *The UMAP Journal* 13: 149-174 (1992).

Answers to Exercises

1. $x = \dfrac{pd}{q+p}$.

2. $L''(x) = p^2(p^2+x^2)^{-\frac{3}{2}} + q^2(q^2+(d-x)^2)^{-\frac{3}{2}} > 0$. Notice that all terms are positive.

3. $T''(x)$ is just $L''(x)$ with $1/c_a$ and $1/c_w$ put in front of the terms.

4. $D''(\alpha) = -4 \cdot \dfrac{d^2\beta}{d\alpha^2}$. Since $\sin\alpha = k\sin\beta$, we have

$$\cos\alpha = k\cos\beta\frac{d\beta}{d\alpha}$$
$$-\sin\alpha = -k\sin\beta\frac{d\beta}{d\alpha} + k\cos\beta\frac{d^2\beta}{d\alpha^2}.$$

Simplification gives $\dfrac{d^2\beta}{d\alpha^2} = \dfrac{\sin(\beta-\alpha)}{k\cos^2\beta} < 0$ since $\beta < \alpha$ and $0 \le (\alpha-\beta) \le 90$ degrees. Hence, $D''(\alpha) > 0$.

5.

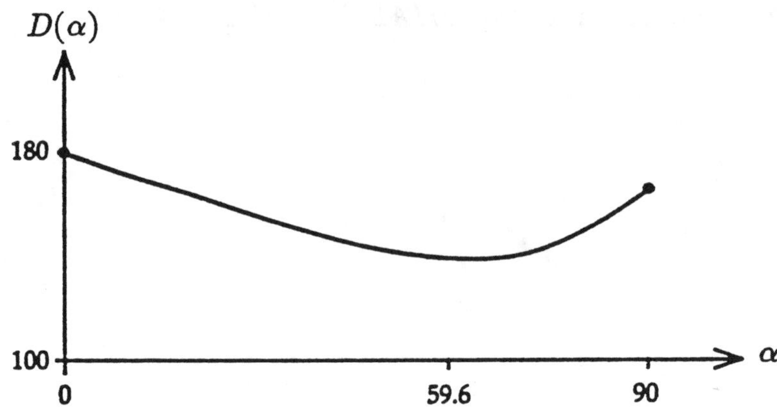

6. Angle of the highest point on the bow plus the angle of the sun equals the rainbow angle 42.5°. Hence the sun is at 17.5°.

7. $D'(\alpha) = 6\dfrac{d\beta}{d\alpha} - 2 = 0$. Also, $\dfrac{d\beta}{d\alpha} = \dfrac{\cos\alpha}{k\cos\beta}$. Hence by squaring,

$$\frac{1}{9} = \frac{\cos^2\alpha}{k^2\cos^2\beta} = \frac{\cos^2\alpha}{k^2 - k^2\sin^2\beta}$$
$$= \frac{\cos^2\alpha}{k^2-\sin^2\alpha} = \frac{\cos^2\alpha}{k^2-1+\cos^2\alpha}.$$

Solving for $\cos\alpha$ gives the result.

8.

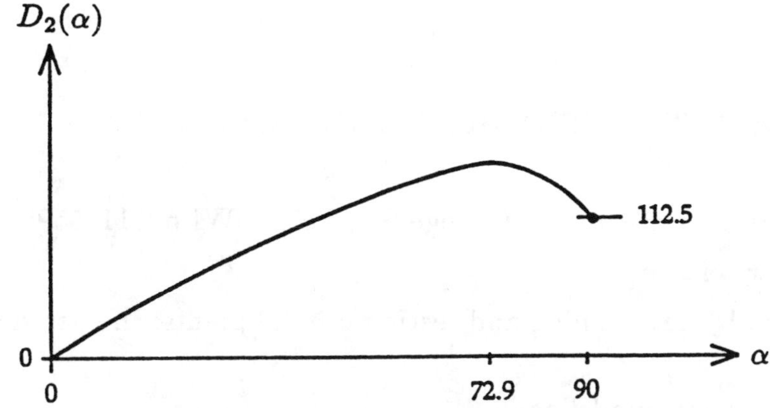

9. Red gives 129.424° and violet gives 126.395°. Violet appears higher in the sky.

10. The exact form of $D_n(\alpha)$ depends on whether you choose to count deflections as being clockwise or counter clockwise. One form gives

$$D_n(\alpha) = 180n + 2\alpha - 2(n+1)\beta.$$

The critical point occurs when $\cos\alpha = \sqrt{\dfrac{k^2-1}{n(n+2)}}$.

THREE OPTIMIZATION PROBLEMS IN COMPUTING

Author: Paul J. Campbell, Beloit College, 700 College St., Beloit, WI 53511–5595

Area of application: computer science

Calculus needed: optimization by determining and testing critical points; the derivative of $\ln x$.

Related mathematics: mathematical modeling

Suggestions for use: the exercises require a scientific-functions calculator

In this module we will consider three optimization problems in the design of computer hardware and software:

- How can a computer disk be formatted to store the maximum amount of data?
- How can data best be stored in "blocks" so as to minimize wasted space?
- How can we minimize time for transferring data between computers, while still checking for accuracy?

We will study the setting of each problem, construct a mathematical model to solve the problem, and compare the results of the theoretical solution to what is actually done. A common theme is that, although the problems are *discrete* (the solutions need to be integers), our models will be *continuous* (allowing real numbers as solutions) so that we can use the techniques of differential calculus.

Maximizing Storage on a Disk

The Problem

As manufactured, computer disks, both floppy diskettes and the platters in hard-disk drives, are just flat surfaces coated uniformly with a magnetic medium. To be used in a particular computer, they must be *formatted* in accordance with that machine's disk operating system. The formatting places little magnetic markers on the disk to divide it into fixed numbers of *sectors* and *tracks*. A sector is analogous to a sector of a circle, and a track is a thin circular ring (see Figure 1). The portion of a track within a single sector is called a *block*; a block is further subdivided into *bytes*, with each byte made up of eight *bits*. Each bit is a single small region that is either magnetized or not, according to whether it represents a 1 or a 0.

The number of tracks per inch is limited by mechanical considerations (how accurately the disk controller can position the read/write head of the disk drive), while the maximum density of bits along a track is limited by magnetic considerations (the need to be able to distinguish two adjacent bits). The density of bits along a track is considerably greater than the density of tracks across the surface of the disk. For design reasons, every block on a disk contains exactly the same number of bytes. Hence each track contains as many bytes as the innermost track.

Our problem is: For the disk to hold as much data as possible, where should the innermost track be located?

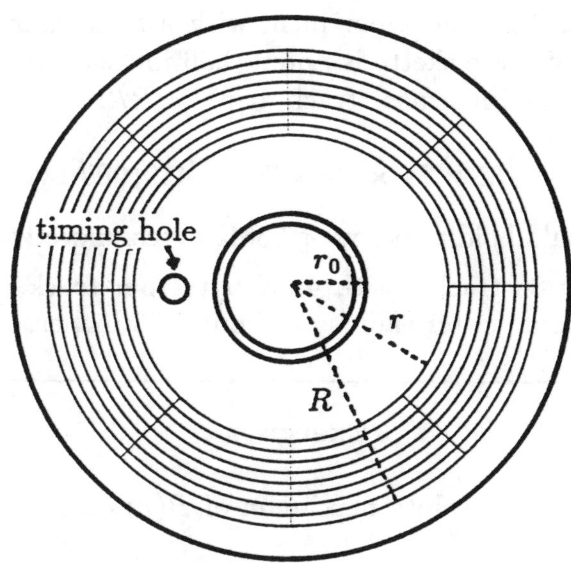

Figure 1. Schematic diagram of organization of data on a disk for which each track has the same number of sectors.

The Mathematical Model

We will build a continuous model for the problem. Suppose the maximum feasible density for tracks is ρ_t tracks per inch (tpi) and the maximum density of bytes along a track is ρ_b bytes per inch. Let the innermost track be at distance r (in inches) from the center of the disk, and let the outermost data track be at (fixed) distance R. Then the number of tracks is $(R-r)\rho_t$.

The innermost track (and hence every track) contains $2\pi r \rho_b$ bytes. Then the total number of bytes that the disk can store is

$$\begin{aligned} B(r) &= (\text{number of tracks}) \times (\text{number of bytes per track}) \\ &= (R-r)\rho_t \times 2\pi r \rho_b \\ &= Cr(R-r), \end{aligned}$$

where C is the constant $2\pi\rho_t\rho_b$. The domain of $B(r)$ is $[r_0, R]$, where r_0 is the distance from the center of the disk to the outer edge of the timing hole (see Figure 1). Using calculus, we find that the maximum of $B(r)$ occurs at $r^* = R/2$ (Exercise 1).

Comparison with Reality

To see how well theory agrees with reality, you can either make measurements on microcomputer diskettes (at the cost of destroying a diskette by opening it up) or else consult the standards for disk formats. We will consider a double-sided, double-density 3.5-inch diskette, which holds 720K bytes (= 720 kilobytes = 720,000 bytes) when formatted for use in an IBM AT-compatible computer.

The formatted disk has 135 tracks per inch, with a maximum density of about 700 bytes per inch. The mylar of the diskette is about 3.36 inches across, so $R \approx 1.68$ inches. If we format the disk with the innermost track at $r^* = R/2 = .84$ inches, then one side holds

$$B = .84 \times 135 \times 2\pi \times .84 \times 700 = 419,000 \text{ bytes}.$$

Hence the two-sided disk could hold as many as 838K bytes, more than the 720K required.

The actual format uses 80 tracks, with 4.5K bytes per track, spaced from an inner radius at about .93 inches to an outer radius at about 1.53 inches.

Exercises

1. Verify that $r^* = R/2$ is an absolute maximum of $B(r)$ on its domain. (Assume that $r_0 < R/2$.)

2. For a high-density double-sided 5.25-inch diskette $R = 2.56$ inches, $\rho_t = 96$ tpi, and $\rho_b = 912$ bytes per inch. What would be its capacity if it were formatted optimally? (It is designed to hold 1200K bytes.)

Optimizing Dynamic Storage

The Problem

Both the main memory of a computer and its secondary storage (usually a disk) are often called upon by the operating system to store data files whose lengths are not known in advance. Ideally, the data should be stored in contiguous memory locations, to enable fastest retrieval of the data. However, a large enough block of contiguous memory may not be available.

One conventional solution to this problem is to store the data in a *linked list*, each of whose nodes has associated to it some fixed amount of storage, a "block", which is the same for each node. Part of the block is used for the address of the next block in the list, and possibly additional information. The rest of a block is available for the data. The data file is broken into chunks, each of which—except the last—fits exactly into the data area of a block, and the data file is stored in blocks corresponding to successive nodes of the linked list (see Figure 2).

Any unused part of the last block is wasted: the larger the block size, the greater potential waste. But if the block size is quite small, then a disproportionate amount of storage is taken up ("wasted") by control information. Our problem is to determine the optimal size for a file storage block, to minimize the average amount of wasted space.

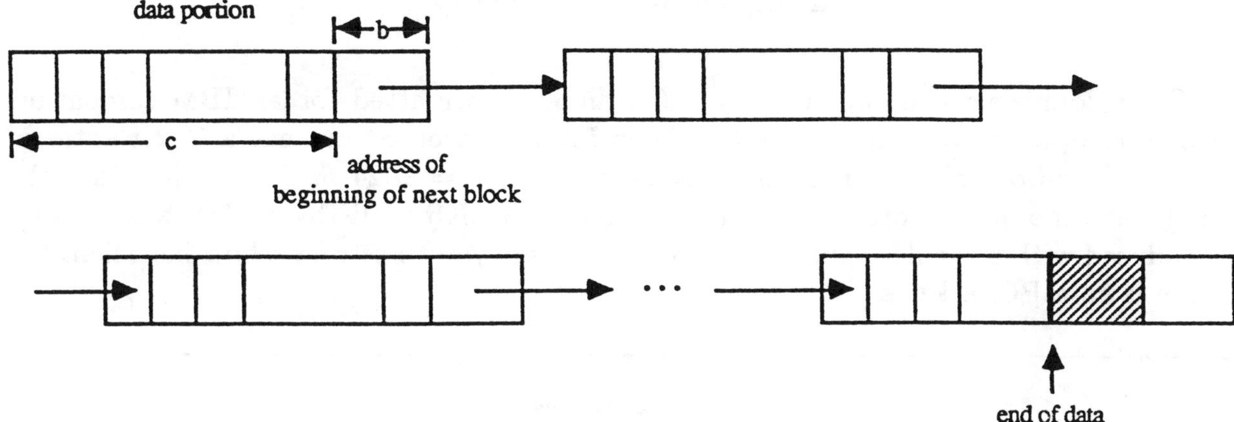

Figure 2. Storage of data in a linked list of blocks.

The Mathematical Model

Following the notation of Wolman [1965], let us take b to be the (fixed) size of the control information in each block, and c to be the size of the data portion of a block. The optimal value of c is to be determined. As in our other models, we will consider c as a continuous quantity, so that we can use calculus in our analysis.

Suppose that on average our data files are L units long. Then we will need

$$\left\lceil \frac{L}{c} \right\rceil$$

blocks, where the brackets denote the *ceiling function*: $\lceil x \rceil$ = the least integer greater than or equal to x. A reasonable assumption is that the average amount of waste in the last block of the data storage is $c/2$. Then the average number of blocks that a file of length L will occupy is just

$$\frac{L}{c} + \frac{1}{2}.$$

Hence the average amount of wasted space is

$$W(c) = \frac{c}{2} + b\left(\frac{L}{c} + \frac{1}{2} + 1\right),$$

where the final "1" counts a segment of b bytes of control information in the directory table that indexes the contents of the entire memory. We find that

$$W(c) = \frac{c}{2} + \frac{bL}{c} + \frac{3b}{2}.$$

Using calculus (Exercise 3), we find that the minimum of W occurs at $c^* = \sqrt{2bL}$.

Comparison with Reality

On a double-sided double-density 720K diskette formatted for an IBM compatible personal computer, each block consists of two 512-byte sectors and has a 12-bit entry in the "file allocation table." In units of bytes, we have $c = 1024$ and $b = 1.5$. Since the average length of a file stored on a diskette is probably 5K to 10K bytes, less waste would be produced with a smaller value for c. Exercise 5 asks you to estimate how inefficient the storage is on a PC diskette.

Exercises

3. Verify that $c^* = \sqrt{2bL}$ is an absolute minimum of $W(c)$ on its domain.

4. Show that if $c = c^*$ and b is small compared with L, then the fraction of storage wasted is approximately $\sqrt{2b/L}$.

5. How inefficient is the storage on a 720K PC diskette? Assume $L = 10,000$ bytes.

Maximizing Throughput on a Noisy Channel

The Problem

Communication within a computer, and between computers, must in many cases be perfect to be effective: if a copy of a program has even one bit miscopied, the copy may not run at all. Some computer memories feature an extra ***check*** (or ***parity***) ***bit*** for each eight-bit byte; the check bit is set to 1 if there is an odd number of 1s in the byte, and to 0 if there is an even number. When the computer processes a byte, it also checks the check bit. If the check bit is not what it should be, then the data have become corrupted—either the check bit itself, or more likely one or more of the bits of the byte, is faulty. We don't know exactly where the defect is, but the data in the byte should not be used.

We will concentrate here, though, on communication by modem between computers. The bits of a file being transmitted are converted by the modem to audible tones (one pitch for a 1, another for a 0). Such communication often takes place over ordinary voice-grade long-distance telephone lines and is subject to "noise."

One way to check whether the data arrive as sent is for the communication to take place under a "protocol" that features some form of error-detection through calculation of a "check" quantity by the receiving computer. If an error is detected, then the part of the file that is affected must be re-sent. How complicated we make the check quantity depends on how sure we want to be to detect any errors. A single check bit will suffice to detect if a single error (or an odd number of errors) has occurred but will miss any even number of errors.

Both checking and resending slow down *throughput*, the volume of data that can be sent per unit time. If we were to wait until the end of the transmission to do a check, we might find that we have to re-transmit the entire file (and conceivably have to repeat it more than once!). On the other hand, sending a check bit with every byte of the file will slow down the rate of transmission of the file in two ways: we have nine-eighths as much to transmit, and the receiving computer must spend time checking each byte and signalling back whether it checks out.

We would like to minimize the average total amount of time needed to send a file, including all re-transmissions of parts in which errors are detected. We will send the data in packets of b bits. After each packet is sent, a check quantity is sent; if an error is detected, the packet is re-sent. We want to determine the optimal size for b.

The Mathematical Model

We assume that the original file consists of N bits, hence is sent as N/b packets (actually $\lceil N/b \rceil$, but we are building a continuous model). Let

$$p = \text{ probability that a bit is received correctly.}$$

If we assume that errors are statistically independent, then the probability that an entire packet is received correctly is p^b, and the probability of a "bad packet" is $1 - p^b$. A result from probability theory says that the average number of packets that need to be sent to achieve N/b successes (error-free packet transmissions) is N/bp^b †. So, on average, transmitting the file will take

$$\frac{N}{bp^b} \times \text{ (time to send a packet, including the check quantity).}$$

We must also include in the time to send a packet the time to send control information (start-of-packet character, packet length, packet sequence number), send the check information, process the check information, and respond whether the check is successful.

† This is the mean of the negative binomial distribution for the the kth success, with $k = N/b$ and probability of success p^k [Larsen and Marx 1986, 222–224].

We will assume that this time does not depend on the length of the packet: that is, we use a single check digit or a small number of check digits, the same regardless of the length b of the packet. With this assumption, the average time to complete a correct transmission of our message is

$$f(b) = \frac{N}{bp^b}\, k(b+\lambda),$$

where k is the time it takes to send one bit and λ is the number of bits in the check quantity. We want to minimize $f(b)$ for values of b between 1 and N, inclusive. Figure 3 shows graphs of f for a typical value $\lambda = 8$ and varying values of p.

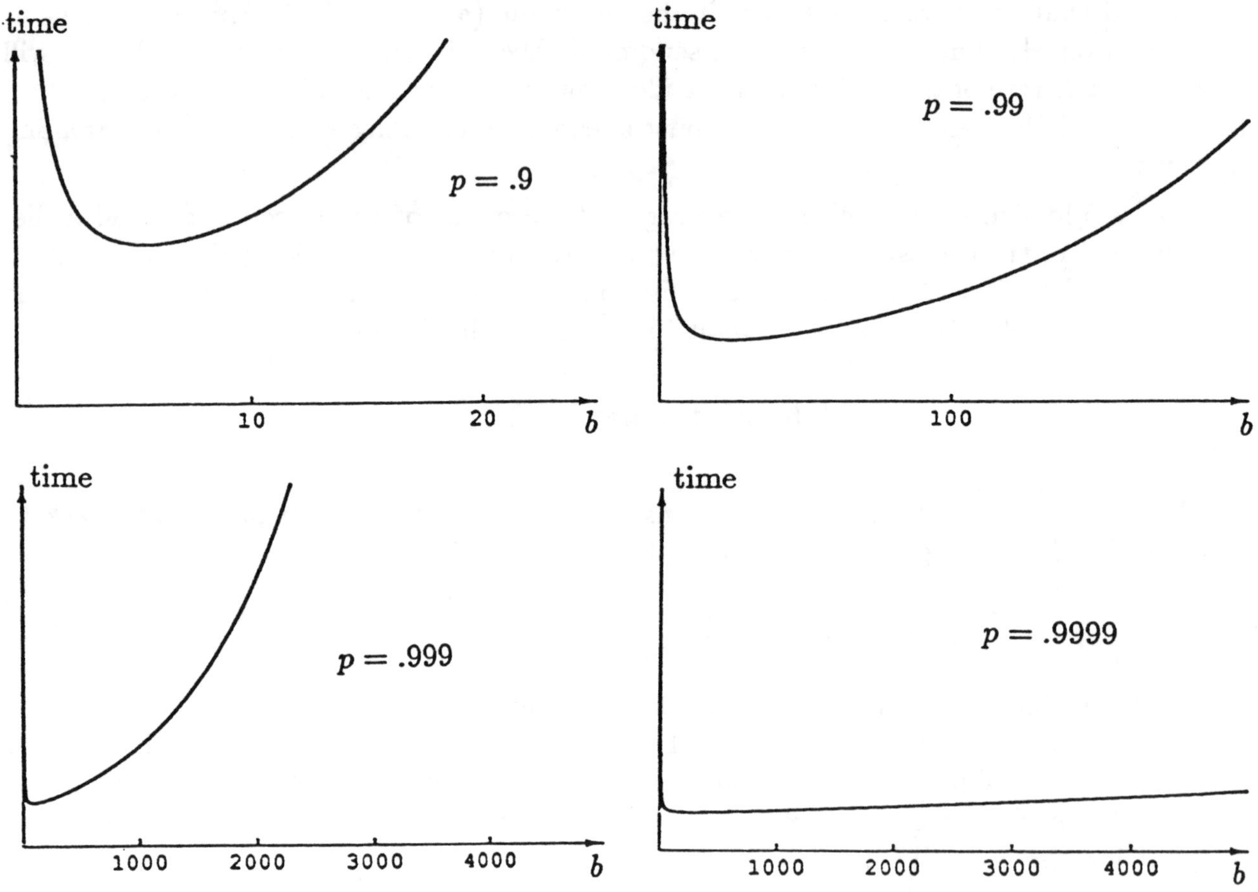

Figure 3. Graphs of $f(b) = \frac{N}{bp^b}k(b+\lambda)$ for $\lambda = 8$ and $p = .9$, $.99$, $.999$, and $.9999$.

We can analyze f for its extreme values by finding its derivative (Exercise 6):

$$f'(b) = \frac{-Nk}{b^2p^b}\left[(\ln p)b^2 + (\lambda \ln p)b + \lambda\right].$$

For the special case $p = 1$ (perfect transmission), we have $\ln p = 0$ and the derivative simplifies to $-Nk\lambda/b^2$, which is always negative; so the extreme values are at the endpoints ($b = N$ to minimize f, and $b = 1$ to maximize f). Otherwise, by using the quadratic formula, we find that the derivative is zero at

$$b^* = \frac{-\lambda \ln p \pm \sqrt{(\lambda \ln p)^2 - 4\lambda \ln p}}{2 \ln p} = \frac{-\lambda}{2} \pm \sqrt{\frac{\lambda^2}{4} - \frac{\lambda}{\ln p}} = \frac{\lambda}{2}\left[-1 \pm \sqrt{1 + \frac{4}{\lambda |\ln p|}}\right]$$

where we have used that $|\ln p| = -\ln p$ since $p < 1$. Under normal circumstances, p will be close to 1, so $|\ln p|$ will be close to 0. Hence $4/\lambda|\ln p|$ will be much larger than 1, and

$$-1 \pm \sqrt{1 + \frac{4}{\lambda|\ln p|}} \approx -1 \pm \sqrt{\frac{4}{\lambda|\ln p|}} = -1 \pm \frac{2}{\sqrt{\lambda|\ln p|}} \approx \pm \frac{2}{\sqrt{\lambda|\ln p|}}.$$

We will be interested in only the positive value. Substituting the approximation into the expression for b^*, we get

$$b^* \approx \sqrt{\frac{\lambda}{|\ln p|}}.$$

To make this formula meaningful, we can examine some numerical instances, as in Table 1. (Note that for values of p near 1, we have $|\ln p| \approx 1 - p$.)

Comparison with Reality

To compare our model with reality, we offer data from the use of the FORMAC protocol for transfers between an IBM mainframe and a Macintosh II [Simware 1989]. The packet size for this protocol is 1,905 bytes, or $8 \times 1905 = 15{,}240$ bits, and it takes about 10 seconds to transfer one such packet over a 2400 bps (bits per second) modem.

We transferred a file of length 105K bytes, which was sent by the protocol as 60 packets (including check information). Of the 60, 35 got through correctly the first time, about a 60% rate. We recorded that 15 of the 25 erroneous packets were re-sent once, 6 were re-sent twice, 1 was re-sent 6 times, 1 was re-sent 15 times, 1 was re-sent 19 times, and 1 was re-sent 23 times. The entire transmission took about 23 minutes.

Table 1. Results for a check function consisting of a λ check bits.

λ	p	$1-p$	$b^* = \sqrt{\lambda/\lvert\ln p\rvert}$	$P = p^{b^*}$
1	.9	.1	3	.729
	.99	.01	10	.904
	.999	.001	32	.968
	.9999	.0001	100	.990
	.99999	.00001	316	.997
	.999999	.000001	1,000	.999
8	.9	.1	9	.387
	.99	.01	28	.755
	.999	.001	89	.915
	.9999	.0001	283	.972
	.99999	.00001	894	.991
	.999999	.000001	2,828	.997
16	.9	.1	12	.282
	.99	.01	40	.669
	.999	.001	126	.882
	.9999	.0001	400	.961
	.99999	.00001	1,265	.987
	.999999	.000001	4,000	.996

Since about 60% of the transmissions got through the first time, we would expect about 60% of the first retries to get through, and 15 of 25 did. We would expect about 60% of the second retries to get through, and 6 of 10 did. But we wouldn't expect needing up to 23 re-transmissions! One explanation may be that the mainframe is heavily burdened at certain times, and transmission attempts at those times suffer long delays. Another explanation is that the communications line may be very noisy for certain intervals, so that errors are more likely and more frequent during those times (so that our assumption that errors are independent does not hold). The result cited above from probability theory shows that, without any overburdening of the mainframe or any noisy intervals, it should have taken on average about $N/bp^b = (N/b)/p^b = 60/0.60 = 100$ packet attempts, or 17 minutes, to transmit the file.

The probability $P = p^b$ of successful transmission of a packet in this case is about 60%; with $b = 15{,}240$, we arrive at the estimate $p = .9999665$. For such a value of p, the optimal packet length is $b^* = \sqrt{\lambda/|\ln p|} \approx 173\sqrt{\lambda}$. We do not know the value of λ that is used by the FORMAC protocol ‡, but we can evaluate the expression for b^* with different possible values of λ. In particular, for $\lambda = 8$ (the common situation of a check *byte*), we

‡ There is no information about it in the manual [Simware 1989], and the vendor did not respond to a letter requesting more information.

get $b^* = 490$; for $\lambda = 1024$—a huge amount of checking—we get $b^* = 5500$. Both of these values are far short of the block size 15,240 used by the protocol.

Sensitivity Analysis

The difference between the optimal packet size and the packet size used by the protocol raises a question relevant in most real-world modeling situations: just how sharp is the optimum? How much does it matter if we use a packet that is twice as long, or half as long, as the optimal size? This is certainly a concern here, as there is no way of knowing in advance how noisy the communications channel will be, that is, the value of p.

In fact, the minimum is fairly broad, as you can see from Figure 3. Exercise 8 asks you do do some relevant calculations. On the other hand, Exercise 8 also shows that in this example the large block size used by the protocol seems to result in significant inefficiency.

Exercises

6. Verify the calculation of $f'(b)$.

7. Use the linear approximation property of the derivative tɔ show that $\ln p \approx p - 1$ for p close to 1.

8. With $p = .9999665$ and $\lambda = 8$ we found that the optimal packet size is $b^* = 490$. Investigate the sharpness of this optimum by computing the ratio $\dfrac{f(b)}{f(b^*)}$ for

   a) $b = 256$ (about half optimal size),

   b) $b = 1024$ (about twice optimal size),

   c) $b = 15240$ (the size used in the protocol).

9. Assuming that the FORMAC protocol uses a single check byte, for what value of p is its block size the optimal block size?

References

Gillett, Philip (1984), *Calculus and Analytic Geometry.* 2nd ed. Lexington, MA: D.C. Heath.

Larsen, Richard J., and Morris L. Marx (1986), *An Introduction to Mathematical Statistics and Its Applications.* 2nd ed. Englewood Cliffs, NJ: Prentice-Hall.

Simware (1989), *Manual for SIM3278.*

Trivedi, Kishor S. (1982), *Probability and Statistics with Reliability, Queuing, and Computer Science Applications.* Englewood Cliffs, NJ: Prentice-Hall.

Wolman, Eric (1965), A fixed optimum cell-size for records of various lengths. *Journal of the Association for Computing Machinery* 12: 53–70.

The section on maximizing storage on a disk was inspired by Gillett [1984, 217–218]. I am indebted to Mark Rogers, technical support engineer for Verbatim, for pointing me to the specific ANSI standard for high-density 5.25-inch disks. The section on dynamic storage arose as an elaboration of a problem in Trivedi [1982, 187, #1], who cites Wolman [1965]. Wolman attributes the simplified model treated in the section to J.B. Kruskal.

Answers to Exercises

1. $B'(r) = C(R - 2r)$, so $r^* = R/2$ is the only critical point. Since $B''(r^*) = -2C < 0$, the critical point is a local maximum. Since it is the only critical point, it must be a global maximum.

2. With $r = 1.28$ inches, one side could hold over 900,000 bytes, so the disk could hold 1800K.

3. We have
$$W'(c) = \frac{1}{2} - \frac{bL}{c^2},$$
so that c^* is the only critical point in the domain $(0, \infty)$. Since $W''(c) = 2bL/c^3$ is positive for $c = c^*$, c^* is the global minimum.

4. $\dfrac{W(c^*)}{L} = \dfrac{1}{L}\left(\dfrac{\sqrt{2bL}}{2} + \dfrac{bL}{\sqrt{2bL}} + \dfrac{3b}{2}\right) \approx \sqrt{2b/L}.$

5. $c^* \approx 173$ bytes, with $W(c^*) \approx 175$ bytes. On the other hand, $W(1024) \approx 529$ bytes. Thus, under our assumption that $L = 10,000$ bytes, the PC diskette wastes about 5% of its space, when it could waste less than 2%.

6.
$$\begin{aligned} f'(b) &= N\left(\frac{-p^b - bp^b \ln p}{(bp^b)^2}\right) k(b+\lambda) + \frac{N}{bp^b}(k) \\ &= Nk\left(\frac{bp^b - (b+\lambda)p^b(1 + b\ln p)}{(bp^b)^2}\right) \\ &= \frac{-Nk}{b^2p^b}\left[(\ln p)b^2 + (\lambda \ln p)b + \lambda\right]. \end{aligned}$$

7. In general, $f(p) \approx f(p_0) + f'(p_0)(p - p_0)$ for p close to p_0. Applying this with $f(p) = \ln p$ and $p_0 = 1$ gives the result.

8. The ratios are a) 1.0068, b) 1.0095, c) 1.6141. While a packet size half or double the optimal size increases the transmission time by less than 1%, the very large size in the protocol increases the transmission time by more than 60%.

9. Solve $b^* = \sqrt{\lambda/|\ln p|}$ for p, getting $p = \exp\{-\lambda/(b^*)^2\}$. For $\lambda = 8$ and $b^* = 15,240$, we get $p = .999999966$.

NEWTON'S METHOD AND FRACTAL PATTERNS

Author: Philip Straffin, Beloit College, Beloit, WI 53511

Area of application: mathematics

Calculus needed: derivatives of polynomials and rational functions, definition of the derivative, the derivative as a linear approximation.

Related mathematics: complex functions, fractals, chaotic dynamics.

Suggestions on use: a computer with appropriate software, or at least a calculator, will be helpful for doing the exercises.

The Problem: How does Newton's Method Behave in the Large?

Some of the oldest problems in mathematics involve finding solutions to equations of the form $f(x) = 0$. Such solutions are called *zeros* of f, or sometimes *roots* of f. For polynomials of degree one or two, general methods of finding zeros were known before 2000 B.C. In the sixteenth century, Italian mathematicians developed methods to find zeros of polynomials of degrees three and four. However, the Norwegian mathematician Niels Abel showed in 1826 that there is no general method for solving polynomial equations of degree greater than four. When f is not a polynomial function, $f(x) = 0$ can be solved exactly only in very special cases.

In cases where we cannot solve $f(x) = 0$ exactly, we need an efficient method of approximating solutions to any desired degree of accuracy. (Indeed, such a method is valuable even when we can solve the equation, but the solution involves, say, calculating cube roots. How can we calculate cube roots efficiently?) Isaac Newton found just such a method, based on his newly developed differential calculus, in 1669. In an improved form due to Joseph Raphson in 1690, this method is now taught in beginning calculus courses as "Newton's method." We will present it in the next section. It involves choosing an "initial guess" x_0, and finding iteratively a sequence of numbers $x_1, x_2, x_3, \ldots$ which converge to a solution.

When the function f has several zeros, the zero found by Newton's method will depend on where we choose the initial guess x_0. The pattern of which initial guesses lead to which zeros—the behavior of Newton's method "in the large"—turns out to be surprisingly complicated and interesting even for polynomials. When we generalize slightly and apply Newton's method to polynomials $f(z)$ where the variable z is a complex number, pictured as a point in the complex plane, the behavior of Newton's method in the large produces pictures which are infinitely complicated and astonishingly beautiful. A sample is shown at the end of this module. Understanding the mathematics behind these pictures is a subject of current research, with strong ties to the study of general chaotic systems. The goal of this module is to guide you along this surprisingly short road from beginning calculus to a research frontier of mathematics.

Newton's Method

Suppose we have a differentiable function $f(x)$, for which we wish to find a zero. We start with an initial point x_0, and determine a new point x_1 by beginning at the point $(x_0, f(x_0))$ on the graph of f and following the tangent line from this point to where it intersects the x-axis. See Figure 1. Since the slope of the tangent line is $f'(x_0)$, we have that

$$\frac{f(x_0)}{x_0 - x_1} = f'(x_0),$$

so that

$$x_1 = x_0 - \frac{f(x_0)}{f'(x_0)}.$$

We then use x_1 as the starting point for the next iteration of this procedure, to get x_2. Thus we generate a sequence of points x_n by the rule

$$x_{n+1} = x_n - \frac{f(x_n)}{f'(x_n)}, \qquad n = 0, 1, 2 \ldots$$

If we choose the initial point x_0 close to the zero x_* we are trying to locate, the x_n's will converge to x_* quite rapidly.

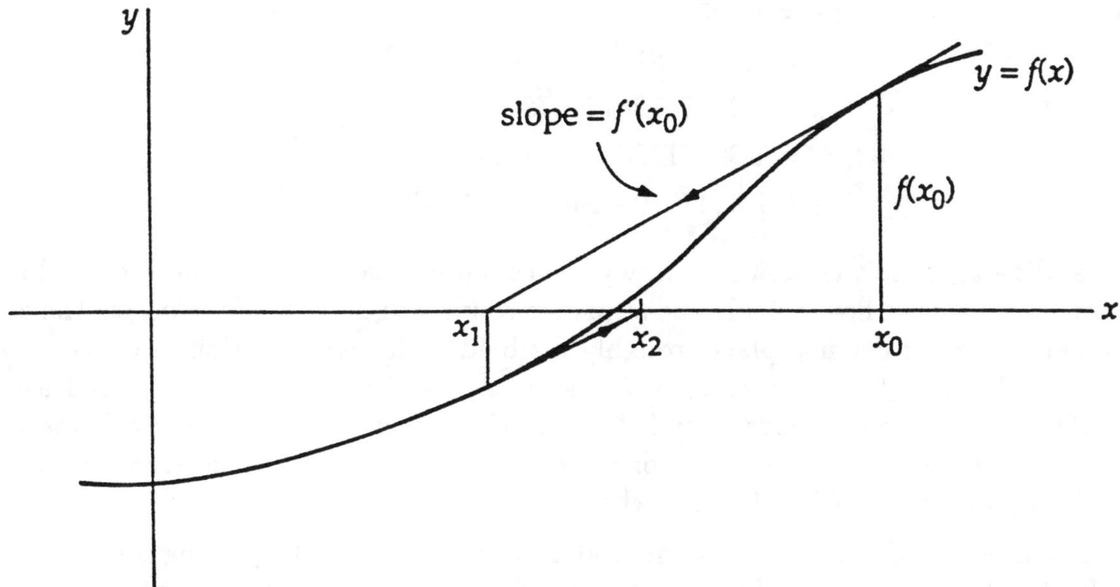

Figure 1. The geometry of Newton's method.

For the purposes of this module, it will be convenient to have a more compact notation for the Newton iteration. Given a differentiable function f, define the *Newton function for* f by

$$N(x) = x - \frac{f(x)}{f'(x)}.$$

Notice that $N(x)$ is not defined if $f'(x) = 0$. Recall that in calculus such a point is called a *critical point* of f. Critical points of f are not in the domain of the Newton function for f.

With this notation $x_1 = N(x_0)$, $x_2 = N(x_1) = N(N(x_0)) = N^2(x_0)$, and in general

$$x_n = N^n(x_0),$$

where the notation N^n means "N applied n times." We will call the sequence $x_0, x_1, x_2 \ldots$ the *Newton sequence* beginning at x_0.

As an example of Newton's method, let us approximate a solution to $x^3 - x - 1 = 0$. A quick graph of $f(x) = x^3 - x - 1$ shows that it has just one zero, between 1 and 2. The Newton function for f is

$$N(x) = x - \frac{x^3 - x - 1}{3x^2 - 1} = \frac{2x^3 + 1}{3x^2 - 1}.$$

Starting with $x_0 = 1$, the results of the Newton calculations are

$$\begin{aligned}
x_0 &= 1 \\
x_1 &= 1.5 \\
x_2 &= 1.34\ldots \\
x_3 &= 1.3252\ldots \\
x_4 &= 1.3247181\ldots \\
x_5 &= 1.324717957244789\ldots \\
x_6 &= 1.324717957244746025960 91\ldots \\
x_7 &= 1.324717957244746025960 91\ldots
\end{aligned}$$

Since x_6 and x_7 agree to the shown accuracy, we conclude that we have found the solution to this accuracy. Notice how quickly the sequence converged once we got close to the zero. The number of correct decimal places roughly doubled with each iteration: x_2 was correct to 1 decimal place, x_3 to 2 places, x_4 to 5 places, x_5 to 13 places, and x_6 to at least 23 places. This kind of convergence is called *quadratic convergence* by numerical analysts, since the error at the $(n+1)^{\text{st}}$ stage is proportional to the square of the error at the nth stage. It is characteristic of Newton's method.

We can understand why Newton's method usually works so well by using some calculus on the Newton function. First of all, from the definition of $N(x)$ we see that x_* is a zero of f if and only if $N(x_*) = x_*$, i.e. x_* is a *fixed point* of N. Next, consider the distance from x_n to the fixed point x_*. If x_n is close to x_*,

$$x_{n+1} - x_* = N(x_n) - N(x_*) \approx N'(x_*)(x_n - x_*).$$

by the linear approximation property of the derivative. Hence if x_n is close to x_*, then x_{n+1} will be even closer if $|N'(x_*)| < 1$. In this case x_* is called an *attracting fixed point* of N. To use this information, we should compute the derivative of N.

By the quotient rule,

$$N'(x) = 1 - \frac{f'(x)f'(x) - f(x)f''(x)}{[f'(x)]^2} = \frac{f(x)f''(x)}{[f'(x)]^2}.$$

Thus, if x_* is a zero of f, we see that $N'(x_*) = 0$. This is certainly less than one, and it means that $x_{n+1} - x_*$ will be much smaller than $x_n - x_*$. In this case, we say that x_* is a *super-attracting fixed point* of N. In fact, with a little closer analysis, it is possible to show that when $N'(x_*) = 0$, $x_{n+1} - x_*$ is approximately proportional to the square of $x_n - x_*$, which is exactly quadratic convergence. What we have said is important enough to summarize in a theorem.

Theorem. *Suppose f is a differentiable function. Then a number x_*, which is not a critical point of f, is a zero of f if and only if it is a super-attracting fixed point of the Newton function of f.*

It is only fair to point out that Newton's method does not always work so beautifully. In fact, for some choices of the initial guess x_0, Newton's method may not converge at all. In the example above, $f(x)$ has critical points at $x = \pm 1/\sqrt{3}$. If we should choose x_0 to be one of those critical points, $N(x_0)$ will be undefined and Newton's method will fail. In fact it will fail if any x_n is a critical point. Finally, there are cases where the entire sequence of x_n's is defined, but does not converge to a zero of f. The simplest case is when the sequence of x_n's becomes *periodic*. For instance, we will see a case shortly where the sequence is periodic with period two: $x_0 = x_2 = x_4 = \ldots$. It is also possible for the Newton sequence to jump about chaotically, never settling down to any regular behavior. Exercises 3 through 6 ask you to think about some of these difficulties. The moral for practical users of Newton's method is clear. *Choose your initial guess wisely, close to where you know there is a zero.*

Exercises

1. Explain why, if $|x_{n+1} - x_*| \approx (x_n - x_*)^2$, the number of correct decimal places in x_n should approximately double with each iteration.
2. For the following functions, calculate the Newton function and iterate it starting at the given initial point. Look for quadratic convergence to a zero of f.
   a. $f(x) = x^2 - 2$; $x_0 = 1$
   b. $f(x) = x^3 - 2x$; $x_0 = 1$
   c. $f(x) = x^3 - 2x$; $x_0 = .7$
   d. $f(x) = x^3 - 2x$; $x_0 = .5$
3. Use the "follow the tangent line" description of Newton's method to explain geometrically why the method fails when $f'(x_n) = 0$ for some n.
4. Sketch the graph of a function f, an initial point x_0, and tangent lines giving x_1 and x_2 with $x_2 = x_0$, so that the Newton sequence is periodic with period two.

5. Calculate the Newton function for $f(x) = \dfrac{x}{x^2+1}$. What happens when you iterate it starting at $x_0 = .5$? What happens when you iterate it starting at $x_0 = 2$? Draw the graph of $f(x)$ and explain these behaviors geometrically.
6. Calculate the Newton function for $f(x) = x^2 + 1$. Iterate it starting at $x_0 = .5$ for as long as you have patience. Do you see any patterns or tendency to converge in the Newton sequence? Why should we expect trouble here?

Newton's Method in the Large: An Example on the Real Line

In Exercise 2b,c,d you saw that when f has several zeros, different initial points x_0 can lead to different zeros. The pattern of which initial points lead to which zeros is not simple. Exercise 2c, for example, shows that Newton's method need not converge to the zero which is closest to x_0. In this section we will study the pattern of convergence to zeros for a slightly simpler polynomial, $f(x) = x^3 - x$. This function clearly has zeros at $x = -1, 0, 1$. Finding the zeros is no problem! Rather, we will be concerned with which initial points lead to which of the zeros.

Definition. *If x_* is a zero of f, the basin of attraction of x_* is the set of all numbers x_0 such that the Newton sequence starting at x_0 converges to x_*. We will denote the basin of attraction of x_* by $B(x_*)$.*

We want to calculate the basins of attraction of $-1, 0$ and 1. Since these points are all attracting fixed points (in fact super-attracting fixed points) of N, we know that some open interval around each one is contained in its basin of attraction. The largest such interval is called the *local basin of attraction.* We will start by finding the local basins.

From the graph of f in Figure 2, it should be clear that if $x_0 \geq 1$, x_n will converge to 1. (Try sketching the first few iterates.) In other words, $[1, \infty)$ is in $B(1)$. Moreover, if x_0 is between the critical point $1/\sqrt{3}$ and 1, it will be true that $x_1 > 1$, so that x_n will still converge to 1. Hence $(1/\sqrt{3}, \infty)$ is in $B(1)$. If $x_0 = 1/\sqrt{3}$, Newton's method fails, so this is the largest open interval about 1 which is contained in $B(1)$: it is the local basin of attraction for 1. Similarly, $(-\infty, -1/\sqrt{3})$ is the local basin of attraction of -1.

Now consider the local basin of attraction of 0. A little experimentation, or a careful look at the graph in Figure 2, shows that close to 0, points oscillate around 0: if $x > 0$ then $N(x) < 0$, and vice versa. This suggests that we might look for a point of period two for N, i.e. a point x for which $N(N(x)) = x$. To do this, we first calculate

$$N(x) = x - \frac{x^3 - x}{3x^2 - 1} = \frac{2x^3}{3x^2 - 1}.$$

Notice that N is an odd function: $N(-x) = -N(x)$. Hence if $N(x) = -x$, then $N(N(x)) = N(-x) = -N(x) = x$, so that x will be a point of period two. Using this observation, we solve

$$-x = \frac{2x^3}{3x^2 - 1}, \qquad 5x^3 - x = 0, \qquad x = 0, \pm\frac{1}{\sqrt{5}}.$$

Hence $\pm 1/\sqrt{5}$ are points of period two for N. Furthermore, it is not too hard to check that if $|x| < 1/\sqrt{5}$, then $|N(x)| < |x|$. This implies that the local basin for 0 is $(-1/\sqrt{5}, 1/\sqrt{5})$. The local basins of attraction are shown in Figure 2.

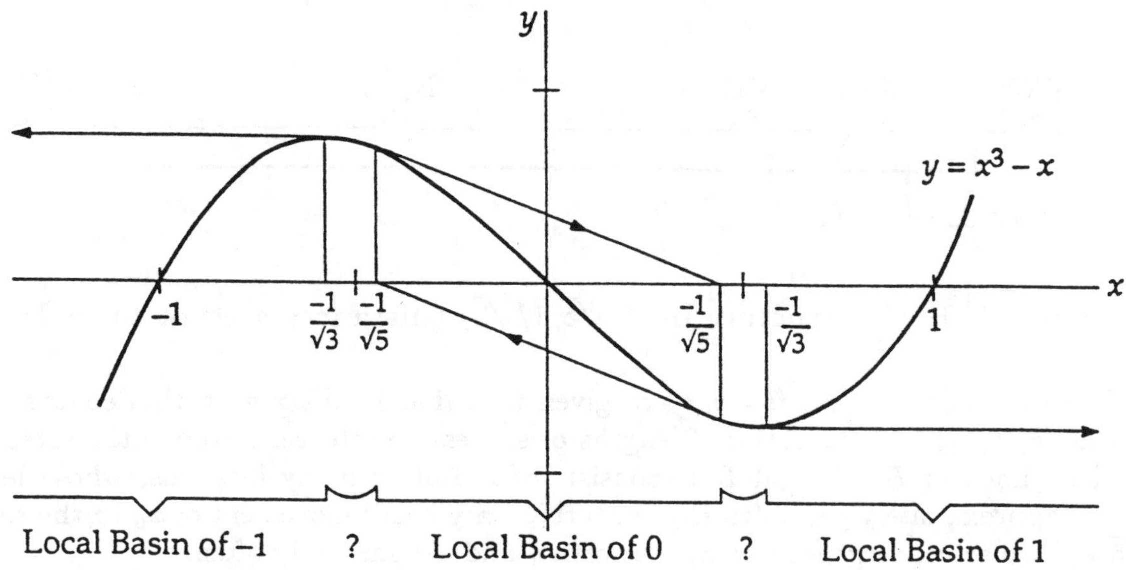

Figure 2. Local basins for Newton's method.

The interesting behavior of Newton's method for this example occurs in the interval $(1/\sqrt{5}, 1/\sqrt{3})$ and the symmetric negative interval. Let's look at these intervals. It should be clear from Figure 2 that if x moves slightly to the left of $1/\sqrt{3}$, $N(x)$ will be large and negative, so that x will be in $B(-1)$. As x continues to decrease , it will stay in $B(-1)$ until $N(x) = -1/\sqrt{3}$. We can solve the equation

$$N(x) = \frac{2x^3}{3x^2 - 1} = -\frac{1}{\sqrt{3}} = -.577350$$

to find $x \approx .465601$. Thus the interval $(.465601, .577350)$ is in $B(-1)$, and by symmetry the interval $(-.577350, -.465601)$ is in $B(1)$.

As x moves to the left of .465601, $N(x)$ moves to the right of $-.577350$ into $B(1)$, so x is in $B(1)$. It stays in $B(1)$ until

$$N(x) = \frac{2x^3}{3x^2 - 1} = -.465601,$$

which happens at $x \approx .450202$. In general, we find a sequence of numbers $b_0 = 1/\sqrt{3} > b_1 \approx .465601 > b_2 \approx .450202 > b_3 > \ldots$ such that

$$(b_i, b_{i-1}) \text{ is in } B(-1) \text{ when } i \text{ is odd,}$$

and

$$(b_i, b_{i-1}) \text{ is in } B(1) \text{ when } i \text{ is even.}$$

The numbers b_i can be determined by successively solving equations $N(b_i) = -b_{i-1}$.

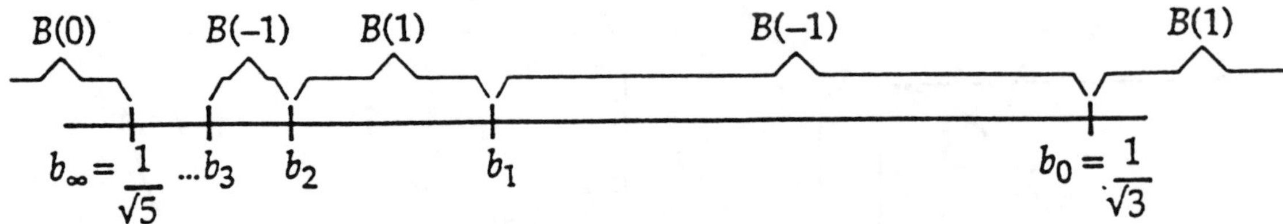

Figure 3. Basin structure in $(1/\sqrt{5}, 1/\sqrt{3})$ (distances are not to scale).

The values of the first few b_i's are given in Table 1, along with the lengths of the intervals (b_i, b_{i-1}) and the ratios of lengths of successive intervals. Notice the interesting behavior. Each of $B(-1)$ and $B(1)$ consists of infinitely many intervals, whose lengths decrease approximately geometrically. An arbitrarily small movement of x_0 to the right of $1/\sqrt{5}$ will cause convergence to shift between 1 and -1 infinitely often.

i	b_i	$b_i - b_{i-1}$	$(b_i - b_{i-1})/(b_{i+1} - b_i)$
0	.577350		
1	.465601	.111749	7.26
2	.4502020	.015399	6.18
3	.4477096	.0024924	6.03
4	.4472962	.0004134	6.01
5	.44722736	.00006884	6.00
6	.44721589	.00001147	6.00
7	.44721398	.00000191	...
...			
∞	.447213595		

Table 1.

Exercises

7. What happens to Newton's method for $f(x) = x^3 - x$ if we choose $x_0 = b_i$ for some i?
8. Verify that $b_1 \approx .465601$ by using Newton's method (!) to approximate the real solution to $2x^3 + .577350(3x^2) - .577350 = 0$. (See where this comes from?)
9. Let's see why the ratios of lengths of successive intervals in Table 1 approach 6.
   a. Verify that $N'(1/\sqrt{5}) = -6$.

From the definition of the derivative, this means that if x' and x'' are any two points close to $1/\sqrt{5}$, we will have

$$\frac{N(x') - N(x'')}{x' - x''} \approx -6.$$

b. Now use the fact that $N(b_i) = -b_{i-1}$, and the fact that b_i is close to $1/\sqrt{5}$ for large i, to show that for large i,

$$\frac{b_i - b_{i-1}}{b_{i+1} - b_i} \approx 6.$$

10. What are the Newton's method basins of attraction for the roots of $x^2 - 2$? What about for the roots of $x^2 - 4x + 3$? Justify your answers.

Newton's Method in the Complex Plane

We have seen that the global behavior of Newton's method can be interesting on the line of real numbers. However, it is in the plane of complex numbers that we see the true intricacy and beauty of the patterns the method can generate. You might want to look ahead at the pictures in this section.

Recall that a complex number z has the form $z = x + iy$, where x and y are real numbers and i is a symbol having the property that $i^2 = -1$. We call x the *real part* of z, and y the *imaginary part* of z. We represent the complex number $z = x + iy$ as the point (x, y) on a coordinate plane which we call the *complex plane.* The x-axis is called the *real axis*; the y-axis the *imaginary axis.* The *norm* of a complex number is the non-negative real number $|z| = \sqrt{x^2 + y^2}$. Geometrically, it is the distance from z to the origin $0 = 0 + 0i$.

Addition and subtraction of complex numbers is done componentwise, so that if $z = x + iy$ and $w = u + iv$, then

$$\begin{aligned} z + w &= (x + u) + i(y + v) \\ z - w &= (x - u) + i(y - v). \end{aligned}$$

Note that $|z - w|$ is the geometric distance between z and w. Multiplication is done using the distributive laws and the property that $i^2 = -1$:

$$zw = (x + iy)(u + iv) = xu + i(xv + yu) + i^2(yv) = (xu - yv) + i(xv + yu).$$

To divide complex numbers, we use the standard method of "rationalizing the denominator":

$$\frac{z}{w} = \frac{x + iy}{u + iv} \cdot \frac{u - iv}{u - iv} = \frac{(xu + yv) + i(yu - xv)}{u^2 + v^2} = \frac{xu + yv}{u^2 + v^2} + i\frac{yu - xv}{u^2 + v^2}.$$

Derivatives of functions of a complex variable behave computationally just like derivatives of functions of a real variable. For example, if $f(z) = z^3 - z$, then $f'(z) = 3z^2 - 1$. Moreover, Newton's method generalizes directly to the complex plane: if $N(z) = z - f(z)/f'(z)$, and z_0 is a complex number, then the iterates $N^n(z_0)$ will in general converge quadratically, in norm, to a zero of $f(z)$.

For a first example of Newton's method in the complex plane, let's consider $f(z) = z^2 + 1$. The corresponding real function $f(x) = x^2 + 1$ has no real roots, and you saw in Exercise 6 that Newton's method responded to this situation by jumping chaotically around the real line. However, $z^2 + 1 = 0$ has two solutions, at $z = i$ and $z = -i$. If we choose z_0 on the real axis, the Newton iterates will do exactly as they did in Exercise 6, since on the real axis, complex arithmetic reduces to real arithmetic. However, if we choose z_0 not on the real axis, Newton's method converges nicely. For example:

$$
\begin{array}{rcrc}
z_0 = & 1 + .5i & z_0 = & .5 - i \\
z_1 = & .1000 + .4500i & z_1 = & .0500 - .9000i \\
z_2 = & -.1853 + 1.2838i & z_2 = & -.0058 - 1.0038i \\
z_3 = & -.0376 + 1.0234i & z_3 = & -i \\
z_4 = & -.0009 + .9996i & & \\
z_5 = & i & &
\end{array}
$$

where the calculations are carried to four decimal places.

If we apply Newton's method to $f(z) = z^3 - z$ with initial points not on the real axis, we notice that global behavior can be delicate and not easily predictable. For example:

$$
\begin{array}{rcrcrc}
z_0 = & .60 + .45i & z_0 = & .65 + .45i & z_0 = & .70 + .45i \\
z_1 = & .4947 + .0222i & z_1 = & .5520 + .0301i & z_1 = & .6067 + .0429i \\
z_2 = & -.8207 - .3249i & z_2 = & -1.3527 - 2.1411i & z_2 = & 1.7051 - 1.7393i \\
z_3 = & -.7866 + .0200i & z_3 = & -.9442 - 1.3527i & z_3 = & 1.1967 - 1.0911i \\
z_4 = & -1.1320 - .0381i & z_4 = & -.6910 - .7892i & z_4 = & .8870 - .6224i \\
z_5 = & -1.0190 - .0103i & z_5 = & -.5437 - .3489i & z_5 = & .7267 - .2445i \\
z_6 = & -1 - .0006i & z_6 = & -.4224 + .1111i & z_6 = & .7689 + .1833i \\
z_7 = & -1 & z_7 = & .0823 - .2777i & z_7 = & .8965 - .1785i \\
 & & z_8 = & .0272 - .0291i & z_8 = & .9371 + .0383i \\
 & & z_9 = & .0001 + .0001i & z_9 = & 1.0034 - .0088i \\
 & & z_{10} = & 0 & z_{10} = & .9999 - .0001i \\
 & & & & z_{11} = & 1
\end{array}
$$

This behavior is shown in Figure 4.

Can we make some sense out of this complicated behavior, and understand the global behavior of Newton's method in the complex plane? In particular, can we find the basins of attraction of the different zeros of $f(z)$, at least for simple functions f? This question was first considered by the English mathematician Arthur Cayley in 1879. Cayley was able to solve the problem completely for quadratic polynomials. He proved the result which you might have guessed in the real case from Exercise 10.

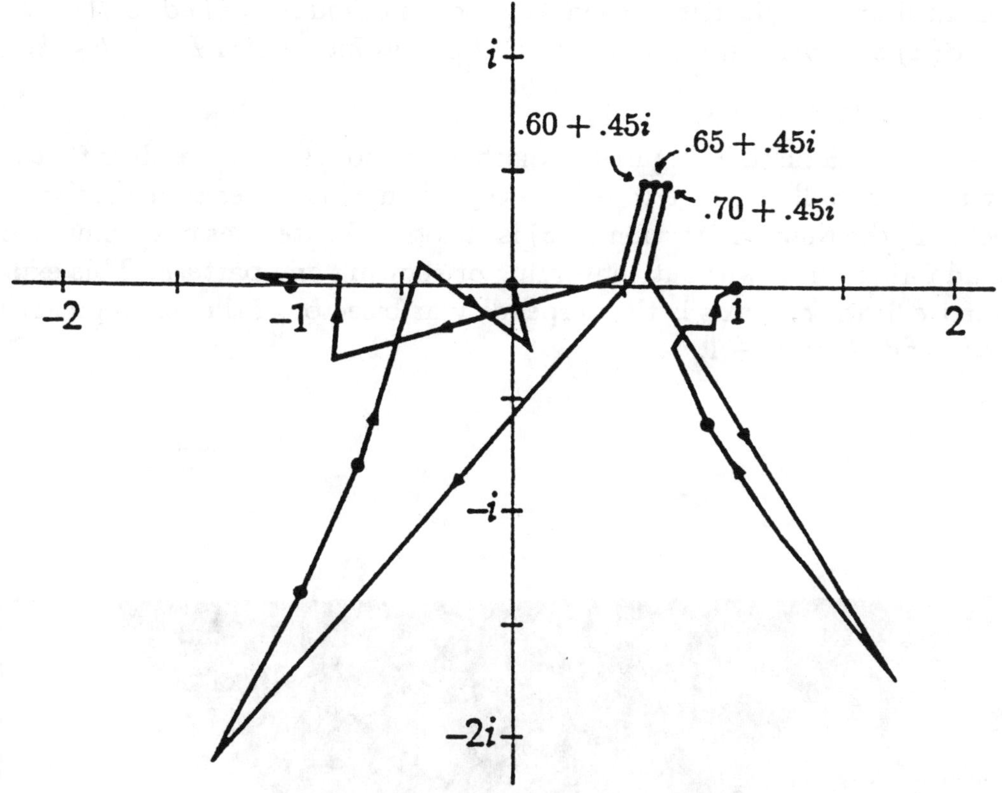

Figure 4. Newton's method for $z^3 - z$.

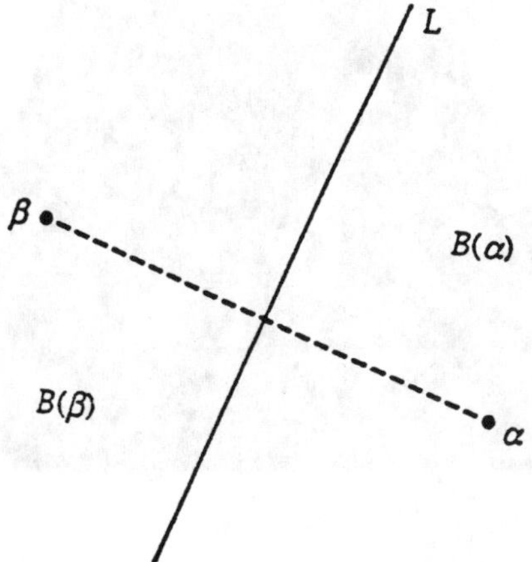

Figure 5. Newton basins for a quadratic polynomial.

Theorem (Cayley, 1879). *Let the complex quadratic polynomial $f(z) = az^2 + bz + c$ have distinct zeros α and β in the complex plane. Let L be the perpendicular bisector of the line segment from α to β. Then, when Newton's method is applied to $f(z)$, the basins of attraction $B(\alpha)$ and $B(\beta)$ are exactly the half-planes into which L divides the complex plane.*

These basins are pictured in Figure 5. Another way to state the result is that Newton's method starting at z_0 will converge to α precisely when z_0 is closer to α than it is to *beta*. On the bisector L, the Newton function $N(z)$ is chaotic: the iterates of a point z_0 on L will bounce around forever on L without converging or showing any pattern. This explains the result of Exercise 6: the real axis is the perpendicular bisector of the line segment between the two zeros i and $-i$ of $z^2 + 1$.

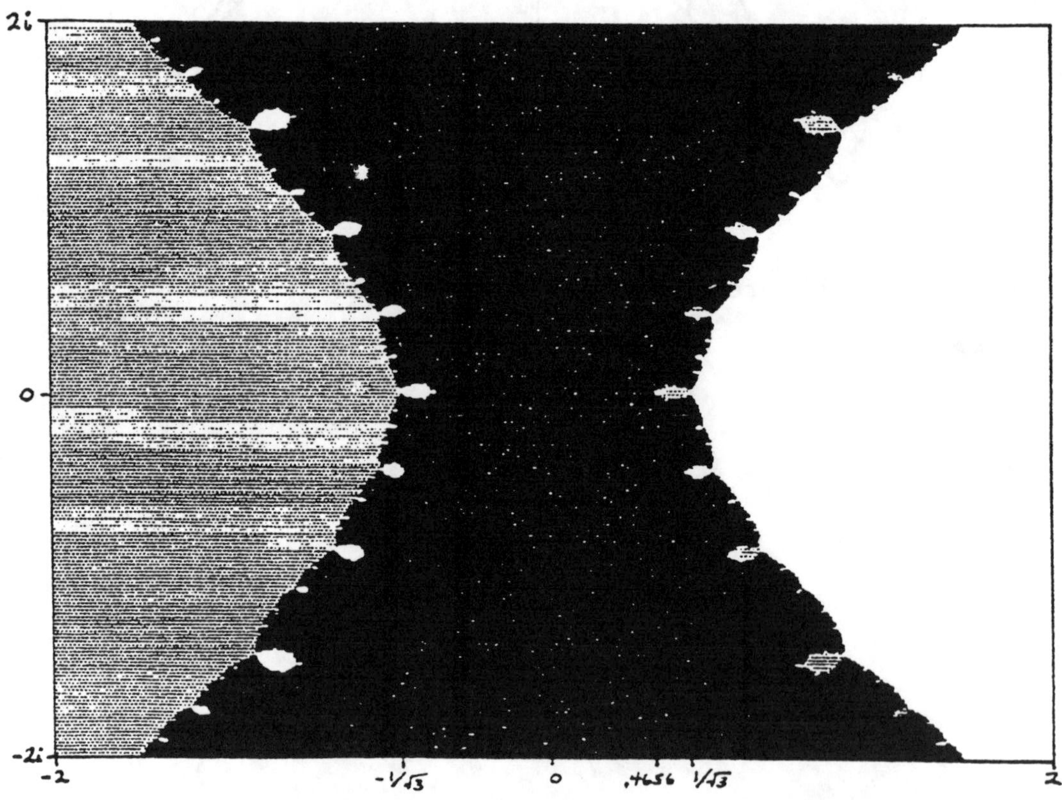

Figure 6. Newton basins for $z^3 - z$.

Our example on the real line shows that the situation for cubic polynomials must be more complicated: initial points will not always converge to the closest zero of f. Cayley considered cubic polynomials, but he wrote that in this case "it is anything but obvious what the division is, and the author has not succeeded in finding it." In fact, even the qualitative nature of the Newton basins for a cubic polynomial wasn't understood until the work of Fatou and Julia on "Julia sets" about 1918, and the first pictures of these basins were drawn only in the 1980's when powerful computer graphics became available.

Let's consider the cubic polynomial $f(z) = z^3 - z$. To picture the global behavior of Newton's method in this example, we ask a computer to color each point z_0 in the complex plane according to which zero of $z^3 - z$ Newton's method will converge to, if we start at z_0. Figure 6 shows the result, with $B(-1)$ colored gray, $B(0)$ colored black, and $B(1)$ colored white. Notice how the pattern we found on the real line extends to an intricate fractal pattern in the complex plane. Figure 7 shows a blow-up of the gray bulb near $z_0 = .5$. Each gray bulb has infinitely many white bulbs attached densely along its boundary, all of those white bulbs have infinitely many gray bulbs attached densely along their boundaries, and so on *ad infinitum*. No wonder Cayley had trouble picturing the shapes of the basins!

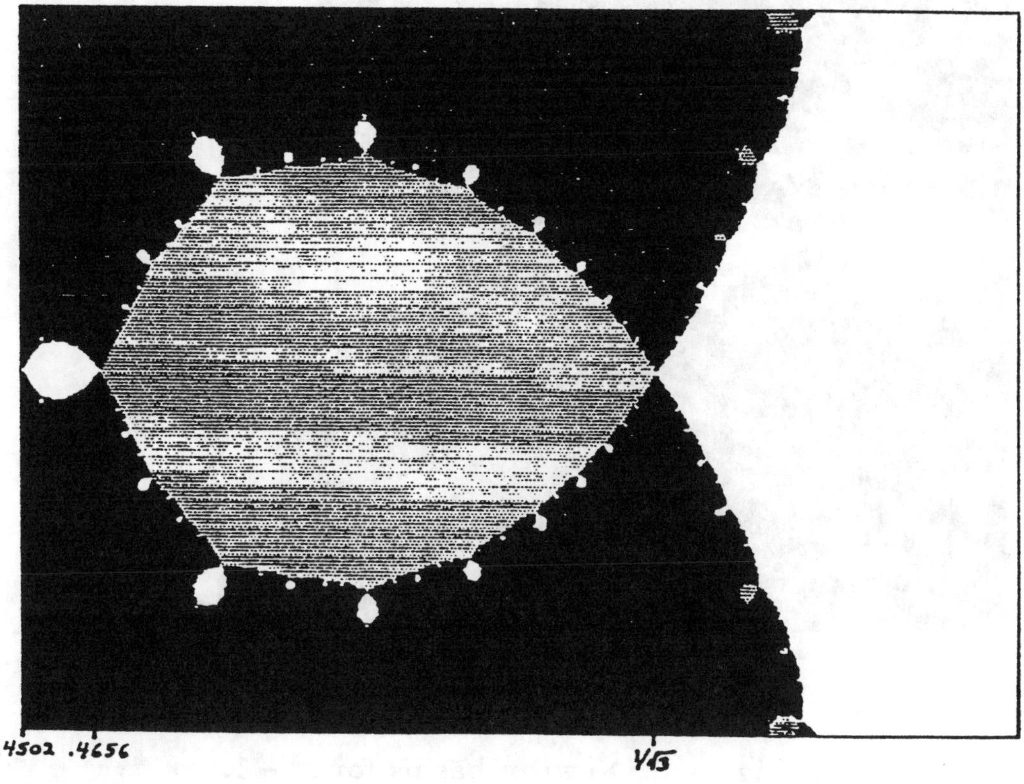

Figure 7. Detail of a bulb.

It turns out that at any point where two of the basins meet, the third basin does as well. The infinitely loopy fractal set which is the common boundary is called a ***Julia set.*** The Newton function $N(z) = 2z^3/(3z^2 - 1)$ takes the Julia set onto itself, and does so chaotically. N also preserves each of the three colored regions, and transforms them in interesting ways. For instance, N takes the gray bulb around $z = .5$ onto the main gray region at the left. N takes three different bulbs onto the bulb around $z = .5$: the bulbs around $z = .6 + .45i$ (see Figure 4), $z = .6 - .45i$ and $z = -.46$ (see Figure 3 and Table 1).

While Figure 6 would probably not win any beauty contests, Newton basins for other cubic and higher order polynomials can have quite beautiful structures. Figure 8 shows part of the basins for Newton's method applied to the cubic polynomial $z^3 - 1$. $B(1)$ is in white, and the other two basins are both colored black. The origin is at the center of the picture.

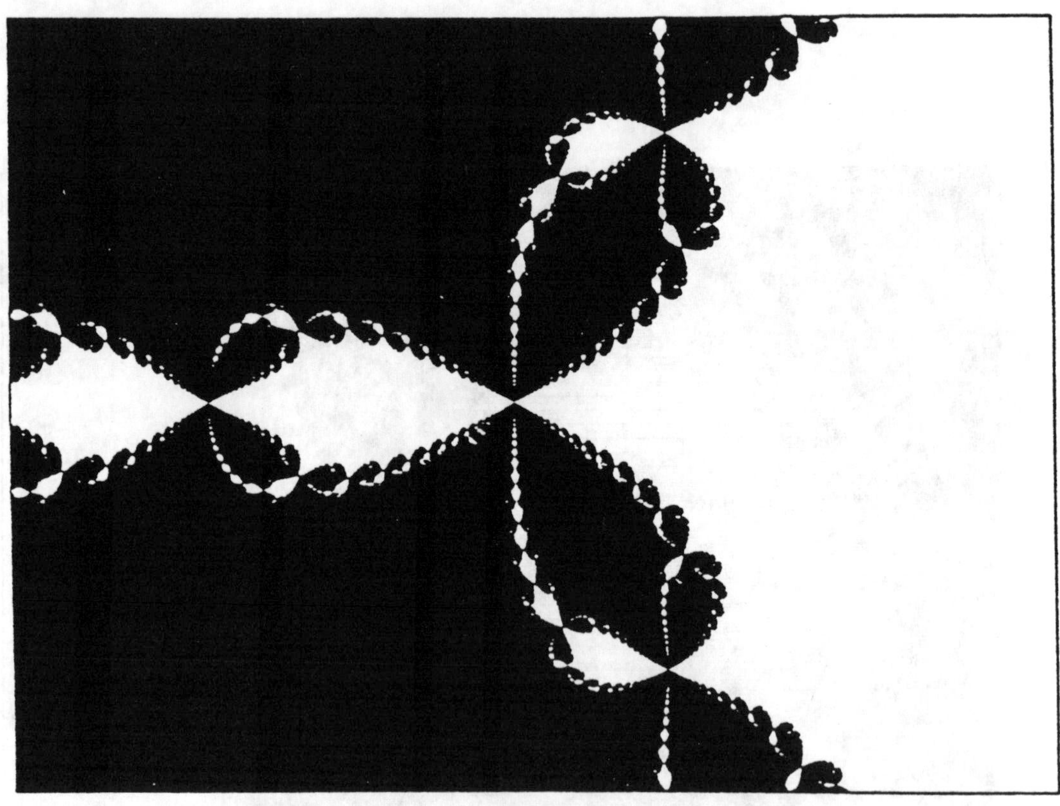

Figure 8. Newton basins for $z^3 - 1$.

Exercises

11. We started our discussion with quadratic polynomials. What about linear polynomials? For $f(z) = az + b$, compute $N(z)$. What is the basin of attraction of the zero $-b/a$?

12. If $f(z) = z^2 + 1$, calculate the first three Newton iterates starting from $z_0 = .5i$. Are they behaving as Cayley's Theorem says they should be?

13. If $f(z) = z^2 + 1$ and $z = x + iy$, find the real and imaginary parts of $N(z)$ as functions of x and y. (This is what you need to do if you are going to program a computer to carry out Newton's method, and your computer does not support complex arithmetic.)

14. Use the formula for the imaginary component in Exercise 13 to show that $N(z)$ maps $B(i)$ (the upper half plane) into itself.

15. If you have a computer, and succeeded well enough in Exercise 13 to want to try something harder, program your computer to do Newton's method on $f(z) = z^3 - z - 1$. Find the three complex zeros of f.

16. Program your computer to do Newton's method on $f(z) = z^3 - 1$ and find three starting values z_0, all with positive real part, which converge to the three different zeros of f. (Figure 8 can help you.)

Further Directions

Chapter 6 of Peitgen and Richter's *The Beauty of Fractals* and the article by Peitgen, Saupe and von Haeseler have a number of other pictures of Newton's method in the complex plane, and there is also a discussion in James Gleick's best-seller *Chaos: Making a New Science.* Becker and Dörfler have a number of do-it-yourself computer experiments involving Newton's method and other ways of generating fractals. Strang shows a number of interesting examples of Newton's method, emphasizing chaotic behavior.

The global behavior of Newton's method in the complex plane is an area of current research in mathematics. For instance, in the examples we have seen, the initial points z_0 for which Newton's method fails to converge to any zero of $f(z)$ are rare. They are points on the the boundary between different basins, and you would have to be very unlucky to choose one by chance. However, in 1983 Curry, Garnett and Sullivan discovered that there are many cubic polynomials for which there are sets of starting values z_0 with *positive area* for which the sequence $N^n(z_0)$ does not converge to any zero of $f(z)$. The structure of these kinds of areas is related to the famous *Mandelbrot set* in fascinating ways.

We seem to have come a long way from an approximation technique based on calculus. Yet, of course, it isn't really a long way at all. Perhaps the moral is that in mathematics, there are simple questions ("What happens if we try different starting points for Newton's Method?") which, if we pursue them, lead to surprising and sometimes beautiful answers and new questions.

References

Becker, Karl-Heinz and Michael Dörfler (1989), *Dynamical Systems and Fractals: Computer Experiments in Pascal*, Cambridge University Press. Chapter 4.

Cayley, Arthur (1879), "The Newton-Fourier imaginary problem," *American Journal of Mathematics* 2: 97.

Curry, James, Lucy Garnett and Dennis Sullivan (1983), "On the iteration of a rational function: computer experiments with Newton's method," *Communications in Mathematical Physics* 91: 267–277.

Gleick, James (1987), *Chaos: Making a New Science*, Viking.

Peitgen, Heinz-Otto, Dietmar Saupe and Fritz von Haeseler (1984), "Cayley's problem and Julia sets," *The Mathematical Intelligencer* 6: 11–20.

Peitgen, Heinz-Otto and Peter Richter (1986), *The Beauty of Fractals*, Springer-Verlag. Chapter 6.

Strang, Gilbert (1991), "A chaotic search for i," *College Mathematics Journal* 22: 3–12.

I am grateful to Brent Halsey for generating Figures 6 and 7. A version of this module appeared in the *UMAP Journal* 12 (1991) 143-164.

Answers to Exercises

1. If $x_n - x_* = c \cdot 10^k$ is correct to k decimal places, then $x_{n+1} - x_* \approx c^2 \cdot 10^{2k}$ will be correct to about $2k$ decimal places.

2. In each case the last figure is the zero correct to the shown accuracy:

1	1	.7	.5
1.5	2	−1.294...	−.2
1.4166...	1.6	−1.43322...	+.0085...
1.4142156...	1.442...	−1.41456...	−.0000006165...
1.4142135623746...	1.4150...	−1.41421370...	.000000000000...
1.414213562373095...	1.4142142...	−1.414213562373...	
	1.414213562373...		

3. The tangent line at $(x_n, f(x_n))$ is level, so never intersects the x-axis to give x_{n+1}.

4. See Figure 2 for an example.

5. $N(x) = \dfrac{2x^3}{x^2-1}$. When $x_0 = .5$, x_n converges to $x_* = 0$. When $x_0 = 2$, x_n goes to infinity.

6. The Newton sequence jumps about chaotically and never converges:

0.5	−0.75	0.29167	−1.56845	−0.46544	0.84153	−0.17339	2.79697
1.21972	0.19993	−2.40088	−0.99218	0.00785	−63.7104	−31.8473	...

We would expect some kind of trouble because $f(x) = 0$ has no real solutions.

7. Newton's method fails because the denominator of $N(x)$ becomes undefined at the $(i+1)^{\text{st}}$ stage.

10. For $x^2 - 2$, $B(-\sqrt{2}) = (-\infty, 0)$ and $B(\sqrt{2}) = (0, \infty)$. For $x^2 - 4x + 3$, $B(1) = (-\infty, 2)$ and $B(3) = (2, \infty)$. Argue from the geometry of the parabola. In the next section we will generalize to the complex plane the principle that for quadratic polynomials, any initial point is attracted to the root closest to it.

11. $N(z) = -b/a$. $B(-b/a)$ is the entire complex plane.

12. $.5i \to 1.25i \to 1.025i \to 1.0003049i \to \ldots \to i$.

13.
$$N(z) = \frac{1}{2}\left(z - \frac{1}{z}\right) = \frac{x}{2}\left(1 - \frac{1}{x^2+y^2}\right) + i\frac{y}{2}\left(1 + \frac{1}{x^2+y^2}\right)$$

14. If $y > 0$, then $\frac{y}{2}\left(1+\frac{1}{x^2+y^2}\right) > 0$.

15. $z = 1.32472$, $z = -.662359 \pm .56228i$. We found the first root in an earlier example.

16. For example, $(1/2) \pm (1/2)i$ lead to $-(1/2) \pm (\sqrt{3}/2)i$.

HOW OLD IS THE EARTH?

Author: Paul J. Campbell, Beloit College, Beloit, WI 53511

Area of application: geology, physics

Calculus needed: solving $y' = ky$.

Related mathematics: mathematical modeling

Suggestions for use: the exercises require a scientific-functions calculator; optional exercises involving least-squares line fitting require a computer program for that purpose.

Controversy over the Age of the Earth

How old is the earth? Estimates of the age of the earth have become larger with time and the uncovering of more evidence. In 1650 Bishop Ussher, basing his calculation on his theology, pinpointed the origin of the earth as occurring in 4004 B.C. Sedimentary rocks—those made up of layers of sediment—he supposed to have resulted from sediments deposited as the Great Flood (of Noah and his ark) receded. Scientists gradually realized that natural processes could not have formed such rocks in so short an interval, and that the earth must be much older. (For a summary of the history of attempts to date the earth, see Badash [1989].)

In the late 19th century, considerations of the hypothesized cooling of the earth led England's leading physicist, Lord Kelvin, to suggest that the earth was 20 to 40 million years old [Burchfield 1975]. But geologists already thought that it must be much older. The discovery of radioactivity at the end of the 19th century at last provided the basis for a reliable method to determine the age of the earth. In this module we explore the method, its mathematical background, and its application to determining the ages of the earth and the moon.

The idea for this module came from two students in Calculus II, who realized that there must be a mathematical explanation behind the graphical procedure they were using in their geology class to determine the age of rocks, and asked me to explore it and explain it to them.

The Relevant Physics

All of the naturally occurring materials in our world are formed from combinations of atoms of 92 basic chemical elements, such as hydrogen, oxygen, carbon, and iron. All of the atoms of a particular element have the same number of protons in their nuclei, so that, for example, every hydrogen atom has 1 proton in its nucleus and every carbon atom has 6 protons in its nucleus.

In addition, an atomic nucleus may include a variable number of neutrons, so that there are different varieties, or *isotopes*, of an element. A hydrogen nucleus, for example, can contain 0, 1, or 2 neutrons, corresponding to ordinary hydrogen, deuterium, and tritium. The notations for these isotopes are ${}^{1}_{1}\text{H}$, ${}^{2}_{1}\text{H}$, and ${}^{3}_{1}\text{H}$, where H is the chemical symbol for hydrogen, the lower number is the number of protons in the nucleus, and the upper number is the total number of *nuclides* (protons plus neutrons) in the nucleus. Similarly, a carbon atom can contain between 3 and 10 neutrons; but only the isotopes with 6, 7, or 8 neutrons occur naturally: ${}^{12}_{6}\text{C}$, ${}^{13}_{6}\text{C}$, and ${}^{14}_{6}\text{C}$, where C is the chemical symbol for carbon. We will omit the lower number when it will simplify notation, since specifying both it and the chemical symbol is handy but redundant.

Different isotopes of the same element differ in weight but have largely the same chemical properties in how they combine with other elements. For example, deuterium (sometimes referred to as "heavy hydrogen") combines with oxygen to form "heavy" water, with properties similar to the ordinary water formed from ordinary hydrogen and oxygen.

Many isotopes are stable but some are not. Unstable isotopes are radioactive: they emit radioactive particles and break down into either a different isotope of the same element or into an isotope of a different element. For example, carbon-14 (${}^{14}_{6}\text{C}$) breaks down into nitrogen (${}^{14}_{7}\text{N}$). The element rubidium (which is especially important in geological dating) has two naturally-occurring isotopes, ${}^{85}_{37}\text{Rb}$, which is stable, and ${}^{87}_{37}\text{Rb}$, which breaks down into strontium-87 (${}^{87}_{38}\text{Sr}$).

For a given unit of time, an atom of a particular unstable isotope has a certain constant probability of breaking down. Whether it does so or not is a random event, unaffected by neighboring atoms. Provided there is a large enough number of atoms in the sample, the statistical law of large numbers guarantees that roughly the same *proportion* of atoms will break down in each time unit. Also, in each of two large-enough collections of atoms of the same radioactive isotope, over a given span of time, the same proportion of atoms will break down. Our calculations will assume that samples are sufficiently large and the time interval sufficiently long for us not to have to worry about chance variation in the proportion of atoms that break down.

Different isotopes break down at different rates. The standard way to measure how fast an isotope breaks down is in terms of its *half-life*, which is the length of time that it takes for one-half of the atoms in a sample to break down. Isotope half-lives vary greatly, from millionths of a second to millions of years.

The Use of Isotopes in Geological Dating

For geologists to be able to date the formation of a rock or mineral, it must have contained, at the time of its forming, an unstable isotope that decays directly into a stable product. Provided we know accurately the half-life of the unstable isotope, we can calculate the age of the rock from

- the amount of stable product generated by decay since the rock was formed, and
- the amount of unstable isotope remaining in the rock now.

Later we will see that quite a few assumptions are involved in our model of this process, including being able to determine these amounts accurately.

For a measurable amount of unstable isotope to remain now in a very old rock, the isotope must have a fairly long half-life. A half-life on the order of a billion years is about right for most geological purposes. Elements such as uranium and rhenium have sufficiently long half-lives to be useful; but carbon-14, with a half-life of 5,730 years, is not useful on a geologic scale of time. Table 1 lists the half-lives of some isotopes that are useful in biological and geological dating. Isotopes of different elements lead to different dating methods; we note the names of some of these in Table 1. (The decay constants in the last column will be discussed shortly.)

Table 1. Half-lives of some isotopes useful in geological and biological dating.

Isotope	Half-life (years)	Dating method	Decay constant (proportion per year)
$^{14}_{6}C$	5,730	carbon-14	1.209×10^{-4}
$^{87}_{37}Rb$	48.6×10^9	rubidium-strontium	1.43×10^{-11}
$^{238}_{92}U$	4.470×10^9	uranium-thorium	1.55125×10^{-10}
$^{234}_{92}U$	248,000	uranium-thorium	2.794×10^{-6}
$^{230}_{90}Th$	75,200	uranium-thorium	9.217×10^{-6}

Exercises

1. Approximately how much remains today of 1,000 grams of uranium-238 formed five billion years ago? ten billion years ago?

2. Approximately how much remains today of 1,000 grams of carbon-14 formed a million years ago? a billion years ago? Hint: First solve the easier problems of how much remains after $1 \times 5,730$ years, after $2 \times 5,730$ years, and $3 \times 5,730$ years, and then generalize.

3. Given an initial amount N of an isotope with a half-life of $T_{1/2}$ years, how much will remain after Y years?

A Mathematical Model of Radioactive Decay

In order to calculate the age of a rock, we need a mathematical model of the decay process of an unstable isotope. At the level of the atom, radioactive decay is a discrete stochastic process, meaning that it deals with individual units subject to chance. The unstable isotope consists of a (large) finite number of atoms, which break down one by one, at random times, into atoms of the stable product.

For calculus to be useful in our analysis, our mathematical model must be

- continuous (not involving individual entities) and
- deterministic (not subject to chance variation).

We will develop a continuous model by assuming that we can deal with the unstable isotope and its stable product, with their large numbers of atoms, as if they were continuous quantities. We will eliminate the role of chance by assuming that the isotope decays continuously (not in bursts), at a constant rate that is directly proportional to the amount of unstable isotope remaining. Thus, exactly half of the remaining quantity decays in each half-life period.

We translate these ideas into mathematical expressions in terms of the following notation, which is common in the literature of isotope geology:

t	time, in years, $t = 0$ being the time of formation of the rock
$N(t)$	the number of atoms of unstable isotope at time t
N_0	the number of atoms of unstable isotope at time $t = 0$
$D(t)$	the number of atoms of stable product at time t
D_0	the number of atoms of stable product at time $t = 0$
$T_{1/2}$	the half-life of the unstable isotope (usually in years)
λ	the *decay constant*, the proportion that decays per unit time (usually one year)

The rate of decay of the unstable isotope is simply dN/dt, and the assumption that the rate of decay is proportional to the amount of unstable isotope remaining becomes the differential equation

$$\frac{dN}{dt} = -\lambda N,$$

in which the constant of proportionality is the negative of the decay constant.

A way to convince yourself that this equation agrees with our definition of the decay constant is to consider what happens in a small interval of time dt. The amount of the unstable isotope that will decay in one year is λN, so the amount that will decay in dt years is $\lambda N\, dt$. Hence the change dN in N is $-\lambda N\, dt$, and this argument from differentials produces the equation above.

We note that our model also has the initial condition

$$N(0) = N_0.$$

Since one atom of unstable isotope breaks down into one atom of stable product, we have

$$D(t) = D_0 + (N_0 - N(t)),$$

or

$$D(t) + N(t) = D_0 + N_0,$$

which is constant throughout the process.

Exercises

4. Show that the solution to the differential equation for $N(t)$, taking into account the initial condition, is
$$N(t) = N_0 e^{-\lambda t}.$$
Sketch the solution curve.

5. Derive the equation for converting between the decay constant and the half-life:
$$\lambda T_{1/2} = \ln 2.$$

6. Write a differential equation and an initial condition for $D(t)$.

7. Use the solution for $N(t)$ from Exercise 4 to solve the equation for $D(t)$ to find that
$$D(t) = D_0 + N_0 \left(1 - e^{-\lambda t}\right),$$
and sketch the corresponding solution curve.

Solving a Simple Version of the Dating Problem

The predictions of the model of radioactive decay in the preceding section agree well with data from experiments. In such experiments, the initial amount of a radioactive substance (or of a stable product) is measured, and then the amount is measured again at a known later time. The decay constant can be determined from the two measurements.

However, our original geological problem is an inverse problem—given the amount present now, how much time has passed since the sample was formed? A moment's reflection will convince you that we can't answer this question just from knowing the amount of the radioactive substance present now. The amount present now could have resulted from a small amount decaying for a small length of time, or from a large amount decaying for a long time. Similarly, knowing just the amount of stable product present now won't do; some or all of it could have been there when the rock was formed.

What will suffice is knowing—in addition to the decay constant—

- the amounts now present of *both* the radioactive substance and the stable product, that is, $N(t)$ and $D(t)$, and
- the initial amount of the stable product, D_0.

Using the last equation in the previous section, we can deduce N_0; and using the results of Exercises 4 or 7, we can find t.

Example. Suppose that 1% of the rubidium-87 in a sample, which originally contained no strontium-87, has decayed to strontium-87. How old is the sample?

We have $\lambda = 1.43 \times 10^{-11}$ (from Table 1), $N(t) = 0.99N_0$, $D_0 = 0$, and $D(t) = 0.01N_0$. From Exercise 4,

$$N(t) = N_0 e^{-\lambda t} = 0.99N_0,$$

$$e^{-1.43\times10^{-11}\times t} = 0.99,$$

$$-1.43 \times 10^{-11} \times t = \ln 0.99 = -0.01005,$$

$$t = 7.0 \times 10^8 = 700 \text{ million years.}$$

Exercises

8. For lead-214, a sample of 1.000 grams decays to 210 milligrams in 1 hour. Determine the decay constant and the half-life in units of minutes.

9. Analysis of a piece of mica shows that 6.00% of the atoms in the rock are rubidium-87 and 0.24% of the atoms are strontium. Of the strontium, 17% is strontium-87. Assuming that all of the strontium-87 was produced by decay of rubidium-87 originally present in the rock, how old is the rock?

10. Use the results of Exercises 4 and 7 to show that

$$\frac{D(t) - D_0}{N(t)} = e^{\lambda t} - 1,$$

and solve for t.

The Isochron Diagram

One major complicating factor in the solution outlined in the exercises of the preceding section is that we cannot be sure that the rock in question did not originally contain, at the time of its formation, some of the stable product.

Another major complication is that it is usually not possible to determine directly the absolute amounts present of the unstable isotope or of the stable product. Instead, mass spectrometer analysis measures relative amounts—the *ratio* of each to another stable isotope.

Let's be specific by considering the *rubidium-strontium method* of dating. Rubidium occurs naturally as ^{87}Rb (unstable) and ^{85}Rb (stable). The radioactive ^{87}Rb decays into ^{87}Sr (stable). Strontium has four naturally-occurring isotopes, all of them stable: ^{84}Sr, ^{86}Sr, ^{87}Sr, and ^{88}Sr. (There is also a short-lived radioactive isotope, ^{90}Sr, which does not occur naturally but is produced in explosions of atomic bombs.) The stable isotope ^{86}Sr provides a good basis for comparisons and measurement of relative amounts of other isotopes, as it is not produced by decay of any naturally-occurring unstable isotope.

Let ^{87}Rb, ^{87}Sr, and ^{86}Sr stand for the quantities present of each isotope. A mass spectrometer can measure the ratios ^{87}Rb/^{86}Sr and ^{87}Sr/^{86}Sr. In our decay model, therefore, it makes sense to take

$$N(t) = \text{ the ratio } \frac{^{87}\text{Rb}}{^{86}\text{Sr}} \text{ at time } t$$

$$D(t) = \text{ the ratio } \frac{^{87}\text{Sr}}{^{86}\text{Sr}} \text{ at time } t.$$

If there were no initial amount of ^{87}Sr present (i.e. $D_0 = 0$), we could proceed as in the example in the previous section. If D_0 were not zero, but were known, we could proceed as in Exercise 10. But how can we incorporate into our calculation an unknown initial concentration D_0 of the stable product?

We can get around this problem for *igneous* rocks. These are formed by cooling and solidification of molten magma. From the same magma, different minerals forming at the same time—in the same *melt*—will contain different proportions of rubidium and strontium. However, the same mixture of strontium isotopes will be present in both minerals; in particular, both minerals have the same initial ratio of strontium-87 to strontium-86, so that D_0 will be the same for both minerals.

Recall from our discussion of the mathematical model for radioactive decay that the combined total number of atoms of the unstable isotope and the stable product is conserved: $D(t) + N(t) = \text{constant}$. We may represent the situation of a mineral on a graph whose axes are labelled with the ratios $D = {}^{87}\text{Sr}/{}^{86}\text{Sr}$ and $N = {}^{87}\text{Rb}/{}^{86}\text{Sr}$. (See Figure 1.)

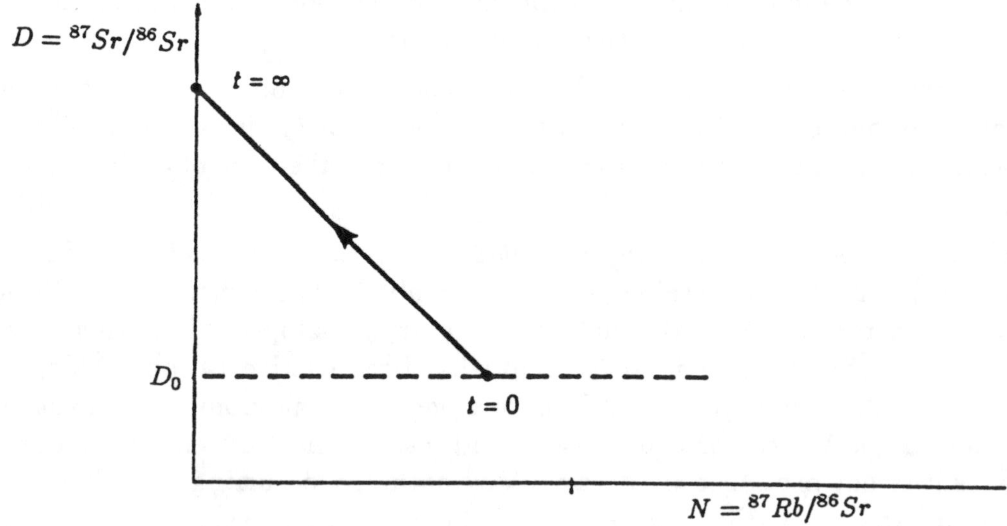

Figure 1. Trajectory of the radioactive decay of a mineral containing rubidium-87.

Since the sum of these ratios stays constant, the point representing the mineral lies on a line of slope -1. As the rubidium-87 decays, the amount of strontium-87 increases; and the point representing the mineral moves upward and to the left. We take $t = 0$ as the time of formation of the mineral. Ultimately, after an infinite amount of time, the point will be on the vertical axis: all the rubidium will have decayed to strontium.

Different minerals from the same melt will start at different points in the N-D plane. Over time, each will trace a straight line segment of slope -1. There are two important features to their graphs. First, all minerals from the same melt begin with the same initial ratio D_0 of strontium-87 to strontium-86, so at time $t = 0$ their points will all be on a horizontal line of height D_0. At $t = \infty$ their points are all on the vertical axis. (See Figure 2.)

The second key feature is that at any time t, $0 < t < \infty$, the points for all minerals from the same melt will lie on the same upward-sloping ray through $(0, D_0)$. In fact, by Exercise 10, for any mineral in the melt, at any time t,

$$\frac{D(t) - D_0}{N(t)} = e^{\lambda t} - 1.$$

This says exactly that the slope of the ray from the point $(0, D_0)$ to $(N(t), D(t))$ has slope $m = e^{\lambda t} - 1$, independent of the mineral. In other words, the points for minerals from the same melt lie on the ray of slope m from $(0, D_0)$. Such a ray is called an *isochron* (from the Greek, "same time"), because the ray passes through points of the same age. Figure

2 shows two isochrons, for $t = t_1$ and $t = t_2$. A graph like Figure 2 showing the isochrons is called an *isochron* (or *correlation*) *diagram*. The isochron diagram provides us with a picture of the quantitative aspects of the decay process.

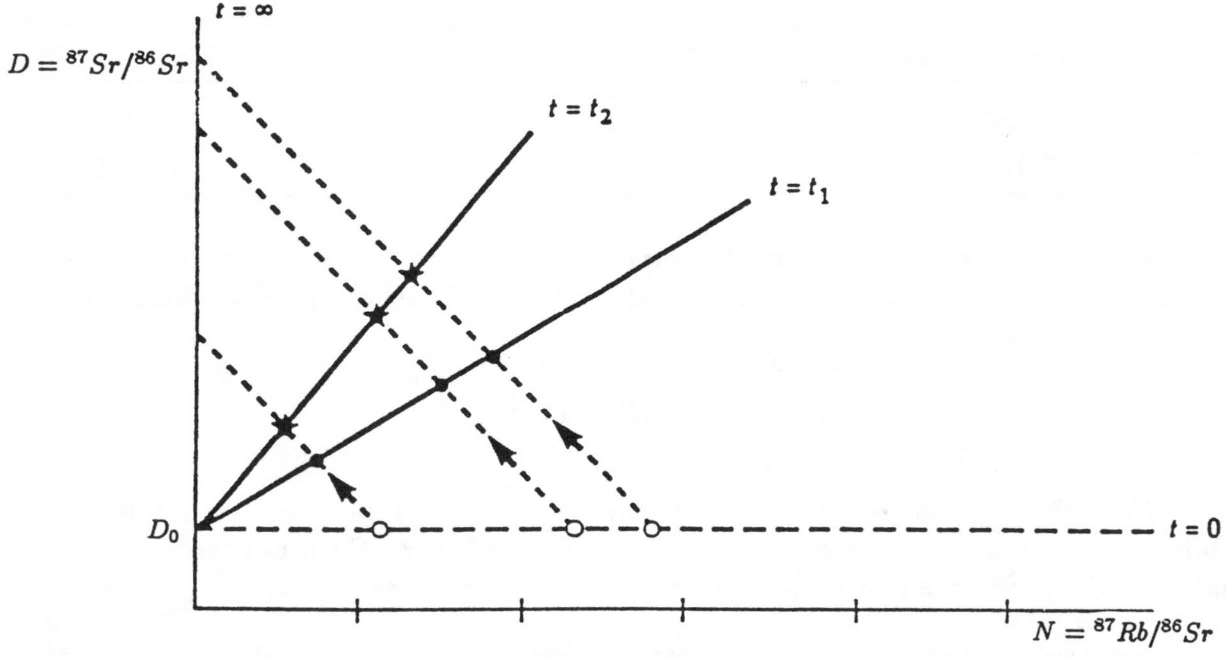

Figure 2. Trajectories for different minerals from the same melt of an igneous rock, with isochrons shown.

Furthermore, and this is the key to dating igneous rocks, we can determine the age of minerals in the rock from the slope of any isochron, since $m = e^{\lambda t} - 1$ implies that

$$t = \frac{\ln(1+m)}{\lambda}. \qquad (*)$$

What we must do is now clear. Take samples of two minerals from the same melt of an igneous rock. use a mass spectrometer to determine

$$\begin{aligned} N_1(t) &= \text{ the current value of } {}^{87}\text{Rb}/{}^{86}\text{Sr for mineral 1} \\ D_1(t) &= \text{ the current value of } {}^{87}\text{Sr}/{}^{86}\text{Sr for mineral 1} \\ N_2(t) &= \text{ the current value of } {}^{87}\text{Rb}/{}^{86}\text{Sr for mineral 2} \\ D_2(t) &= \text{ the current value of } {}^{87}\text{Sr}/{}^{86}\text{Sr for mineral 2.} \end{aligned}$$

Calculate the slope of the current isochrone

$$m = \frac{D_1(t) - D_2(t)}{N_1(t) - N_2(t)},$$

and the solve for the age of the rock containing the minerals using formula $(*)$.

Example [Rapp et al. 1967, 247, 276]. A hypothetical rock containing the minerals feldspar, biotite, and muscovite was analyzed and produced the hypothetical data in Table 2. How long ago was this rock formed?

Table 2. Analysis of hypothetical rock.
(from Rapp et al. [1967, page 247])

Isotope ratio	Feldspar	Biotite	Muscovite
$^{87}Sr/^{86}Sr$	0.77	0.80	0.82
$^{87}Rb/^{86}Sr$	2.00	5.00	7.00

Since we do not directly know the original $^{87}Sr/^{86}Sr$ ratio, we must determine m by using two minerals. Using any two, we find $m = 0.01$. Then we can determine the age of the rock by

$$t = \frac{\ln(1+m)}{\lambda} = \frac{\ln 1.01}{1.43 \times 10^{-11}} \approx 7.0 \times 10^8.$$

The rock is about 700 million years old.

(Note that the three data points in this example lie *exactly* on the isochron, so it didn't matter which pair of points we used to determine m. Such are the advantages of artificial data! See the Exercises for similar problems with real data.)

Exercises

11. Table 3 gives real data for various minerals from an ancient granite. Calculate the points for the isochron diagram for the four minerals in this rock. Why is it difficult to plot the isochron? Use the data for biotite and apatite to determine the age of the rock. Why is it a good idea to use those two minerals, rather than another pair?

Table 3. Rock analysis of an ancient granite.
(adapted from Hamilton [1968, page 441])

Mineral	$\frac{^{87}Rb}{^{86}Sr}$	$\frac{^{87}Sr}{^{86}Sr}$
Biotite	658.6	10.36
Muscovite	14.0	0.91
Feldspar	13.4	0.86
Apatite	0.07	0.71

12. (optional) For the isochron diagram in the previous exercise, use the slope of the least-squares line of best fit to the data to determine the age of the rock.

13. (adapted from Faure [1986]) Four samples of rocks from Northern Light Lake, Ontario, were analyzed by Hanson et al. [1971] with the resulting data in Table 4.

Table 4. Analysis of whole-rock samples from Northern Light Lake, Ontario.
(from Faure [1986, page 176])

Sample	$\frac{^{87}Rb}{^{86}Sr}$	$\frac{^{87}Sr}{^{86}Sr}$
21	0.1564	0.7068
22	0.0755	0.7037
23	0.216	0.7091
24	0.328	0.7133

Plot an isochron diagram for this rock. Use the data for Samples 22 and 24 to determine the age of the rock.

14. (optional) For the isochron diagram in the previous exercise, use the slope of the least-squares line of best fit to the data to determine the age of the rocks.

Critique: Limitations of the Dating Process

Accurate dating of rocks or other objects by comparing isotope ratios rests on several key assumptions. First, we have the **model assumptions**:

• The isotope ratio has changed only by radioactive decay. The relative amounts of the isotopes involved have not changed for any other reason, such as metamorphism (physical change) of the rock, exposure to air or other potential contaminant, or *fractionation* (migration) of atoms within the rock. All of these events are distinctly possible under some conditions.

• The decay constant has not changed over time. This appears to be a very sound assumption for the long-lived isotopes used in dating. The decay constants of some elements increase slightly under very high pressure, but there is no evidence that elements in the crust of the earth have been subjected to such pressures for any length of time [Faure 1986, page 41].

Second, we have the **sampling assumption**:

• The samples analyzed are representative of the rock, mineral, or formation to be dated.

In addition, the accuracy of the dating is affected by **measurement errors** in determining:

• the decay constant (or equivalently, the half-life) of the unstable isotope. The value is known to three or more significant digits for some isotopes. For rubidium-87, the decay constant is known with an uncertainty of about 2% [Dalrymple, 1991].

• the amounts and relative quantities of the isotopes present in the sample. "Isotope ratios [of the same element] can be measured routinely with a precision of ±0.05% or better, and quantities of Rb and Sr to ±1 to 2%" [Long and Giletti 1972, p. 1052]. The main exception is for very young or very small samples. Over a short interval of time—short, that is, compared to the half-life of the unstable isotope—only a very small quantity of the stable isotope is formed by decay. Such a small quantity may be difficult to measure accurately, particularly if it is swamped by a large amount of the stable isotope present initially.

The major source of uncertainty in our calculated value for t, therefore, comes from the uncertainties of 1% to 2% in the amounts of strontium and rubidium and hence in the values of the ratio ^{87}Rb/^{86}Sr that are used in calculating m. Except in cases where samples are small, we can expect the value of t to have an uncertainty of 1% to 2%. Using a greater number of samples and the least-squares to find m can reduce this uncertainty.

How Old Is the Earth? And What about the Moon?

The techniques we have explored in this Module are part of *geochronology*, the study of the ages of rocks. Their main applications are the establishment of an accurate timeline for fossils and extending the timeline beyond the earliest fossils.

So how far back do fossils and rocks go? The oldest known fossils, from the Warrawoona Group in Western Australia, date from 3.3 to 3.5 billion years ago. An intriguing isotopic argument suggests that there may have been life on earth even earlier. There are two stable isotopes of carbon, carbon-12 and carbon-13; but plants preferentially use carbon-12 for building organic compounds in photosynthesis. Thus, sediments with high ratios of carbon-12 to carbon-13 may have been formed from material that had once been part of early plants. Such sediments have been found in rocks from western Greenland which were deposited about 3.8 billion years ago. The oldest dated rocks are 3.96 billion years old. The oldest dated minerals are 4.2-billion-year-old zircon crystals from Australia, for which we do not know the source rock.

The Apollo 17 astronauts brought back from the moon samples of surface soil containing orange glass. Scientists initially anticipated that this material represented recent volcanic eruptions, but the glass turned out to be unexpectedly ancient. Husain and Schaeffer [1973] determined its age by comparing ratios of isotopes of argon, arriving at 3.7 billion years. Various isotopic methods lead to the conclusions that the earth, the moon, and most meteorites formed about 4.5 billion years ago.

Exercises

15. Moorbath et al. [1972] performed rubidium-strontium age determinations of rocks in West Greenland, finding that the rocks were older than any other terrestrial rocks dated up to the time of their study. For 25 rocks from the Narssaq area, their isochron gives an initial $^{87}Sr/^{86}Sr$ ratio of 0.7015 ± 0.00008. This information determines the vertical intercept of the isochron line. They do not give the slope of the isochron, but we can get a good approximation by calculating the slope from the vertical intercept to one other point lying very close to the isochron line. Such a point is their sample 125523, which has a $^{87}Rb/^{86}Sr$ ratio of 2.630 and a $^{87}Sr/^{86}Sr$ ratio of 0.8424. From this information, calculate the age of the rocks.

16. (optional) Here are the complete data from Moorbath et al. [1972, 80] for rocks from the Qîlangarssuit area:

Sample No.	$^{87}Rb/^{86}Sr$	$^{87}Sr/^{86}Sr$
155733	0.200	0.7109
155735	1.926	0.8065
155736	1.053	0.7545
155737	0.223	0.7129
155739	3.262	0.8739
155740	2.612	0.8374
110870	0.802	0.7462

Plot these points on an isochron diagram. Use the slope of the least-squares line of best fit to the data to determine the age of the rocks.

References

Badash, Lawrence (1989), "The age-of-the-earth debate," *Scientific American* 261 (August): 90–95.

Burchfield, Joe D. (1975), *Lord Kelvin and the Age of the Earth*, Science History Publications.

Dalrymple, G. Brent (1991), *The Age of the Earth*, Stanford University Press.

Elliot, Mark, and Nick Lord (1990), "Radioactive decay series," *Mathematical Gazette* 74: 163–167.

Faure, Gunter (1986), *Principles of Isotope Geology*, 2nd ed., Wiley.

Faure, Gunter, and J.L. Powell (1972), *Strontium Isotope Geology*, Springer-Verlag.

Hamilton, E.I. (1968), "The isotopic composition of strontium applied to the problem of the origin of the alkaline rocks," in *Radiometric Dating for Geologists*, edited by E.I. Hamilton and R.M. Farquhar, 437–463, Wiley.

Hanson, G.N., *et al.* (1971), "Age of the early Precambrian rocks of the Saganaga Lake–Northern Light Lake area, Ontario–Minnesota," *Canadian Journal of Earth Sciences* 8: 1110-1124.

Horelick, Brindell, and Sinan Koont (1989), "Radioactive chains: parents and children," Modules in Undergraduate Mathematics and Its Applications 234, *The UMAP Journal* 10: 217–235.

Horgan, John (1991), "In the beginning..." *Scientific American* 264 (February): 116–125.

Hurley, Patrick M. (1959), *How Old Is the Earth?*, Doubleday.

Husain, Liaquat, and Oliver A. Schaeffer (1973), "Lunar volcanism: age of the glass in the Apollo 17 orange soil," *Science* 180: 1358–1360.

Long, Leon E., and Bruno J. Giletti (1972), "Rubidium-strontium dating method," in *Encyclopedia of Earth Sciences*, vol. IVA: *Encyclopedia of Geochemistry and Environmental Sciences*, edited by Rhodes W. Fairbridge, 1052–1056, Van Nostrand Reinhold.

Moorbath, S., R.K. O'Nions, R.J. Pankhurst, N.H. Gale, and V.R. McGregor (1972), *Nature: Physical Science* 240: 78–82.

Rapp, George *et al.* (1967),"Measurement of geologic age of rocks," *Journal of Geological Education* 15: 217-278.

Smith, David G., ed. (1982), *The Cambridge Encyclopedia of Earth Sciences*, Cambridge University Press.

York, D., and R.M. Farquhar (1972), *The Earth's Age and Geochronology*, Penguin.

A longer version of this module, which also discusses uranium-thorium and carbon-14 dating methods, appeared in *The UMAP Journal* 13: 43-78 (1992). I wish to thank Amy Ollendorf '83 and Marjorie Snope Perlin '83 for their curiosity, and various members of the Beloit College faculty for commenting on early drafts of this Module and rendering other assistance: Carl Mendelson for answering my geological questions, sharing with me his teaching materials, and pointing me to the book by Faure; Rama Viswanathan for details on mass spectrometry; and Phil Straffin for editing.

Answers to Exercises

1. The table shows that the half-life of uranium-238 is 4.470×10^9 years. After five billion years, slightly less than one-half of the original amount will remain; after ten billion years, slightly less than one-fourth.

2. After $n \times 5,730$ years, $\left(\frac{1}{2}\right)^n \times 1,000$ grams remain. Since 1,000,000/5,730 = 174.5, the amount remaining is less than $\left(\frac{1}{2}\right)^{174} \times 1,000 = 4.2 \times 10^{-50}$ grams and more than $\left(\frac{1}{2}\right)^{175} \times 1,000 = 2.1 \times 10^{-50}$ grams. For comparison, there are 6.02×10^{-23} atoms in 14 grams of carbon-14. So we can be sure that there is no measurable amount of carbon-14 remaining after a million years, much less after a billion years! Hence, the dating of traces of organisms that lived that long ago must be based on elements other than carbon.

3. $$N\left(\frac{1}{2}\right)^{Y/T_{1/2}}.$$

6. $dD/dt = \lambda N(t) = \lambda N_0 e^{-\lambda t}, \quad D(0) = D_0.$

7. 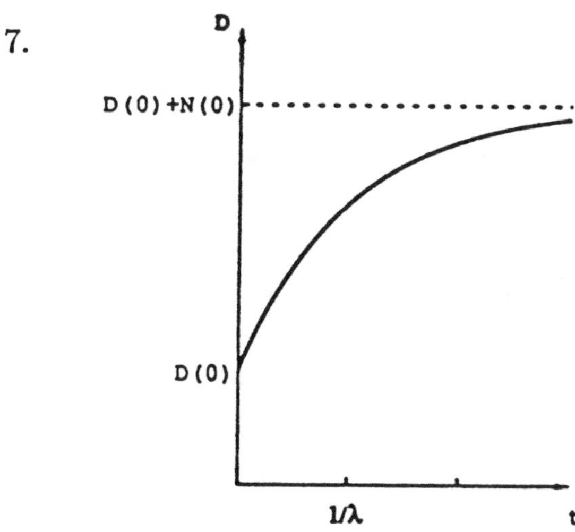

8. $N_0 = 1.000$ grams, $N(60) = 0.210$ grams, $N(t) = N_0 e^{-\lambda t}$. Solving, we find $\lambda = 0.026$. From Exercise 5, $T_{1/2} = \ln 2/\lambda = 26.6$ minutes.

9. $4.7 \times 10^8 = 470$ million years.

10.

$$t = \frac{1}{\lambda} \ln \left(1 + \frac{D(t) - D_0}{N(t)} \right).$$

11. The four data points are (658.6,10.36,) (biotite), (14.0, 0.91) (muscovite), (13.4, 0.86) (feldspar), and (0.07, 0.71) (apatite). The isochron diagram is difficult to plot because the point for biotite is so far from the others. The other points are relatively close together, so using a pair of them is unlikely to lead to a reliable line. The biotite and apatite points are the two points farthest apart. They lead to an age of the rock of 1 billion years.

12. $m = 0.0147$, giving an age of 1 billion years. The biotite point exerts a great deal of influence on the slope; the slope for the least-square line for the other three points is $m = 0.01294$, leading to an age of 900 million years.

13. $m = 0.038$, leading to an age of 2.6 billion years.

14. $m = 0.03803$, leading to an age of 2.6 billion years.

15. 3.65 billion years.

16. $m = 0.05291$, leading to an age of 3.6 billion years.

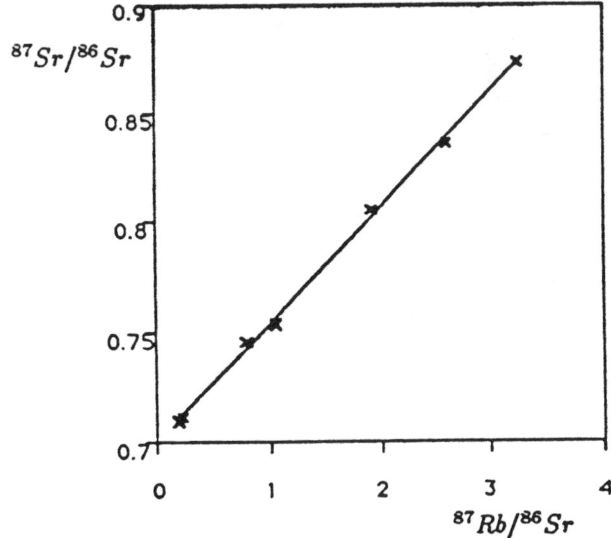

FALLING RAINDROPS

Author: Walter J. Meyer, Adelphi University, Garden City, NY 11530

Area of Application: physics.

Calculus needed: exposure to differential equations. For certain exercises, integrating to solve separable differential equations.

Related mathematics: mathematical modeling.

The Problem: How Fast does a Raindrop Fall?

Our experience tells us some things about how raindrops fall. For instance, large raindrops during a thunderstorm fall faster than small raindrops during a drizzle. The tiny raindrops in fog seem not to be falling at all. Even the largest raindrops do not fall fast enough to do us damage when we walk in the rain.

If we want to understand more closely how objects fall, we build mathematical models. Although a mathematical model can be powerful and helpful, it usually describes only limited aspects of the phenomenon we are trying to understand; hence it will give predictions which are only approximately correct. Often the limitations and approximations are acceptable for our purposes. If they are not, we can sometimes build a new and better model by adding certain refinements. In this module we illustrate these ideas by examining the limitations of Galileo's model of gravity. We study a number of replacements. Each of these alternatives is useful in some circumstances but inappropriate in others.

Our notation for studying falling bodies will be as follows. We use t to measure time, in seconds, and x to measure distance, in feet. The distance axis is vertical and its positive direction points toward the earth, so that as a body falls x increases. Since x is a function of t, it will often be written $x(t)$. Likewise, the velocity of a body v varies with time and we may denote it $v(t)$.

Model 1 (Galileo's Model)

Historians of science often assert that the first truly modern scientific mind belonged to a brilliant (but allegedly cantankerous) Italian professor of mathematics named Galileo (1564–1642). He devised the first good mathematical model of gravity, based on the following assumption:

> An object, falling to the earth from a *moderate height* and *subjected only to the force of gravity*, gains an extra 32 feet/second in velocity for each second in which it falls. In other words, its acceleration is a constant value of 32 feet/second^2. This is true no matter what the weight of the object is.

The italicized phrases are escape clauses, and we shall soon see their significance.

We can express Galileo's assumption with any one of the following equivalent equations:

$$v(t) = v(0) + 32t \tag{1}$$

$$\frac{dx}{dt} = v(0) + 32t \tag{2}$$

$$\frac{d^2x}{dt^2} = 32 \tag{3}$$

where $v(0)$ denotes the initial velocity.

Example 1

A golf ball is at rest at $t = 0$ and is dropped to earth at that instant. The experimenter times its fall and discovers that it takes 1.5 seconds to reach the earth. How fast is it going when it lands? In this problem $v(0) = 0$. Therefore, (1) implies

$$v(1.5) = 0 + (32)(1.5) = 48 \text{ feet/second.}$$

Can we check the prediction of this example experimentally? It's possible but not easy because measuring the velocity of a golf ball when it hits the earth is difficult. Fortunately, we can derive some consequences of our model which are easier to test. Integrating both sides of (2) gives

$$x(t) = v(0)t + 16t^2 + c, \tag{4}$$

where c is a constant. To evaluate c, substitute $t = 0$ and obtain $c = x(0)$, i.e., the distance traveled after 0 seconds. This is 0. So $c = 0$, and Equation (4) becomes

$$x(t) = v(0)t + 16t^2. \tag{4'}$$

Example 2

How far has the golf ball of Example 1 traveled after 1.5 seconds? Using Equation (4′) and recalling that $v(0) = 0$, we immediately obtain $x(1.5) = 16(1.5)^2 = 36$ feet. This conclusion is testable with relatively simple equipment—a tape measure and stopwatch. A fancier way to do this experiment involves flash photography. Beginning when the ball is dropped, the camera shutter clicks at equal time intervals, but always on the same frame of the film. The various positions of the ball are frozen on the same photograph. Figure 1 shows a drawing of what such a photo might look like. Measurements on the photo can be converted to actual distances by multiplying by a suitable scaling factor. Table 1 shows data collected and scaled this way from a photograph where the camera clicked every $\frac{1}{30}$ second. If you convert centimeters to feet (2.54 centimeters = 1 inch), you will see that these data agree pretty well with model 1.

Table 1

Ball position	Distance fallen, centimeters
1	7.70
2	16.45
3	26.25
4	37.10
5	49.09
6	62.18
7	76.36
8	91.58
9	107.89
10	125.34

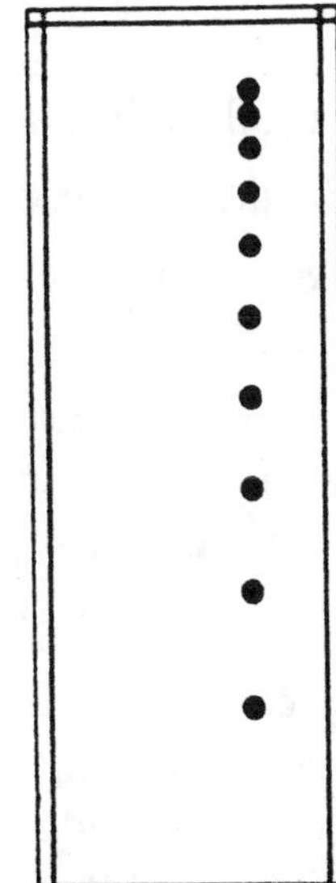

Figure 1

The next example, however, shows an experiment we could do that would contradict model 1.

Example 3

A raindrop, beginning at rest, falls from a cloud 1024 feet above the ground. How long does it take to reach the ground? If we set $x(t) = 1024$ in Equation (4′) and solve for t, we obtain

$$1024 = 16t^2$$
$$t = 8 \text{ seconds.}$$

However, if we actually performed the experiment on a number of raindrops, we would discover two things that contradict model 1: first, that the weight of the raindrop makes an important difference in the time it takes to fall and, second, that the fastest time (for the largest raindrop) is about 40 seconds—5 times as long as predicted by model 1 in the above calculation.

What has gone wrong? The problem is that model 1 is only valid if the object is subjected only to the force of gravity. (Recall that this was one of our escape clauses.) In the case of raindrops, the force of gravity is opposed by a significant amount of air drag—a lucky thing for us or we might be killed by falling raindrops. Air drag is also present when we drop a golf ball, but it is smaller in relation to the force of gravity because of the greater density of the golf ball and the shorter distance of the fall.

The previous example shows that we need a new model to cope with raindrops. Actually we will develop two in detail: one for very small drops and another for large ones.

Exercises

1.a. How far will a golf ball, starting from rest, fall in 5 seconds?

b. How many seconds will it take a golf ball to fall 64 feet if it starts from rest?

c. How many seconds will it take a golf ball, starting from rest, to attain a velocity of 160 feet/second?

2.a. Suppose an object, beginning with a velocity of 0 at $t = 0$, falls in accordance with model 1. How fast is it going after falling 16 feet? 100 feet? [*Hint*: First find the time elapsed. Use Equations (4′) and (1).]

b. Find a formula expressing $v(t)$ as a function of $x(t)$ for objects obeying model 1.

3. Table 1 contains observed values. Write a computer program to compute the theoretical distances according to model 1. The method is shown in Example 2. Use the program to compute distances at multiples of $\frac{1}{30}$ second and compare them to the actual values in the table. (Be sure to convert your theoretical values to centimeters for comparison with the table.)

Model 2 (Stokes' Law)

For spherical drops falling in motionless air and having a diameter $D \leq 0.00025$ feet, the acceleration g due to gravity (the 32 in Equation (3)) is opposed by an amount proportional to the velocity of the raindrop, specifically by an amount equal to $(C/D^2)dx/dt$, where C is an experimentally determined constant, equal to 0.329×10^{-5} feet^2/second. Thus,

$$\frac{d^2x}{dt^2} = g - \frac{C}{D^2}\frac{dx}{dt}, \qquad C \approx 0.329 \times 10^{-5}. \tag{5}$$

Instead of trying to solve this differential equation (see Exercise 5 for a method of solving it), we shall show that it predicts something drastically different from the predictions of model 1: the existence of a *terminal velocity*, that is, an upper bound which the velocity approaches ever more closely as time proceeds.

Theorem 1. *Model 2 implies a terminal velocity* v_{term}.

Proof: By setting the right side of Equation (5) equal to zero, we discover that $d^2x/dt^2 = 0$ when $dx/dt = (g/C)D^2$. If the drop ever achieved this velocity, then d^2x/dt^2, the rate of change of the velocity, would be zero and the body would continue at this velocity. As long as $dx/dt < (g/C)D^2$, d^2x/dt^2 is positive, which means that the velocity increases toward $(g/C)D^2$. Hence

$$v_{term} = \frac{g}{C}D^2.$$

Although a drop falling according to equation (5) never quite reaches its terminal velocity, its velocity eventually becomes so close to v_{term} that, for practical purposes, it is equal to v_{term}. Furthermore, clouds are sufficiently high and a raindrop gets close to its terminal velocity so quickly that it is not a bad assumption to suppose that the drop travels at its terminal velocity for its whole trip. See Exercise 6.

Example 4

Find the terminal velocity of a drizzle drop with diameter $D = 0.00025$ feet. Compare it to the terminal velocity of a fog droplet with one-tenth of that diameter ($D = 0.000025$). We substitute the value of D in the expression for v_{term}. For g, we will use the slightly more accurate value of 32.2 feet/second^2. For the drizzle drop,

$$v_{term} = \frac{32.2(0.00025)^2}{0.329} \times 10^5 = 0.612 \text{ feet/second},$$

a bit more than 7 inches/second. For the fog droplet,

$$v_{term} = \frac{32.2(0.000025)^2}{0.329} \times 10^5 = 0.00612 \text{ feet/second}.$$

This is one one-hundredth the velocity of the drizzle drop. At this slow rate, drops need about 165 seconds to fall 1 foot. This, of course, corresponds exactly to our experience of fog: it hardly seems to be falling at all; mostly it just appears to hang around. Indeed we often notice fog lifting. This is because its rate of fall is so slow that the slightest updraft will overcome it.

Exercises

4. Calculate terminal velocities for drops with the following diameters: $D = 0.00005$, 0.00010, 0.00015, and 0.00020. Use these calculations to make a graph of v_{term} plotted against diameter. What kind of curve do these points lie on?

5. In model 2 (Stokes' law), if we set $v = dx/dt$, the differential equation becomes $dv/dt = g - \frac{C}{D^2}v$. Integrating, we get

$$\int \frac{dv}{g - \frac{C}{D^2}v} = \int dt.$$

a. Perform the integrations and show that $v = \frac{gD^2}{C}(1 - e^{-Ct/D^2})$. In evaluating the constant of integration, assume $v = 0$ when $t = 0$.

b. An approximation sometimes used for e^x, when x is small, is $e^x \approx 1 + x$. Substitute this in the formula found in (a) and show that model 1 results.

c. Replace v by dx/dt in the formula found in (a) and then solve for x as a function of t.

6.a. Let $t \to \infty$ in the formula found in Exercise 5a to find a formula for v_{term}. How does it compare with the one in the text? Show that we may write the formula for v as $v/v_{term} = 1 - e^{-Ct/D^2}$.

b. Use the formula for v/v_{term} to calculate the time at which a raindrop with diameter 0.00025 feet reaches 99 percent of its terminal velocity.

c. Repeat part (b) for diameter 0.000025 feet.

Model 3 (Velocity-Squared Model)

In order to study the effects of gravity on larger raindrops, we now introduce another model. For spherical raindrops falling in still air and having diameter $D \geq 0.004$ feet, the acceleration due to gravity is opposed by an amount proportional to the square of its velocity, specifically an amount equal to $(k/D)(dx/dt)^2$, where k is an experimentally determined dimensionless constant equal to 0.00046. Thus Equation (3) is replaced by

$$\frac{d^2x}{dt^2} = g - \frac{k}{D}\left(\frac{dx}{dt}\right)^2, \qquad k \approx 0.00046. \tag{6}$$

This model, like model 2, predicts a terminal velocity which we can obtain by setting $d^2x/dt^2 = 0$ and solving for dx/dt. The result is

$$v_{term} = \sqrt{\frac{gD}{k}}. \tag{7}$$

Example 5

For a raindrop of diameter 0.004 feet, find the terminal velocity. Also, find how long it takes to reach the ground if it starts its descent in a cloud 3000 feet high. From (7), $v_{term} = \sqrt{\dfrac{32.2 \times .004}{.00046}} = 16.7$ feet/second. Observe how much faster this is than the terminal velocities of the drizzle drop and fog droplet in Example 4.

To find how long it takes the drop to fall 3000 feet, we make the assumption that it reaches its terminal velocity nearly instantaneously. If we suppose it goes at 16.7 feet/second for the whole distance, the time is $3000/16.7 = 180$ seconds, or 3 minutes.

Exercises

7. Calculate terminal velocities of raindrops with the following diameters: 0.005, 0.006, 0.007, 0.008, 0.009, and 0.010. Use these calculations to make a graph of v_{term} plotted against diameter. What kind of curve do these points lie on?

8. Set $v = dx/dt$ in the differential equation for model 3 and show by integration that

$$\frac{dx}{dt} = v = \sqrt{\frac{gD}{k}}\,\frac{e^{bt} - 1}{e^{bt} + 1},$$

where $b = 2\sqrt{kg/D}$. Use the initial condition $v = 0$ when $t = 0$ to evaluate the constant of integration.

9.a. Let $t \to \infty$ in the formula for v found in Exercise 8 to find a formula for v_{term}. Use this to find a formula for v/v_{term}.

b. Use the formula for v/v_{term} to find the time at which v is 99 percent of v_{term}, if $D = .004$ feet.

Model 4 (General Air-Drag Model)

The reader may have noticed that although we have given models for small ($D \leq 0.00025$) and big ($D \geq 0.004$) drops, we have given no model for medium-sized ($0.00025 < D < 0.004$) drops. There is a more general model, which applies to drops of any diameter and incorporates both models 2 and 3. However, as we shall see, the general model is inconvenient to use. Here is a brief description of it.

For a spherical raindrop of diameter D the differential equation describing the fall function $x(t)$ is

$$\frac{d^2x}{dt^2} = g - \frac{0.00092}{D}F(D\,dx/dt)\,(dx/dt)^2. \tag{8}$$

The notation $F(D\,dx/dt)$ means that F is a function of the product $D\,dx/dt$. We will call this experimentally determined function the *drag coefficient function.* A glance at Figure 2 shows that the function is not a simple one which could be expressed by a convenient formula. Note that the figure actually shows the graph of $\log F$ plotted against $\log(D\,dx/dt)$.

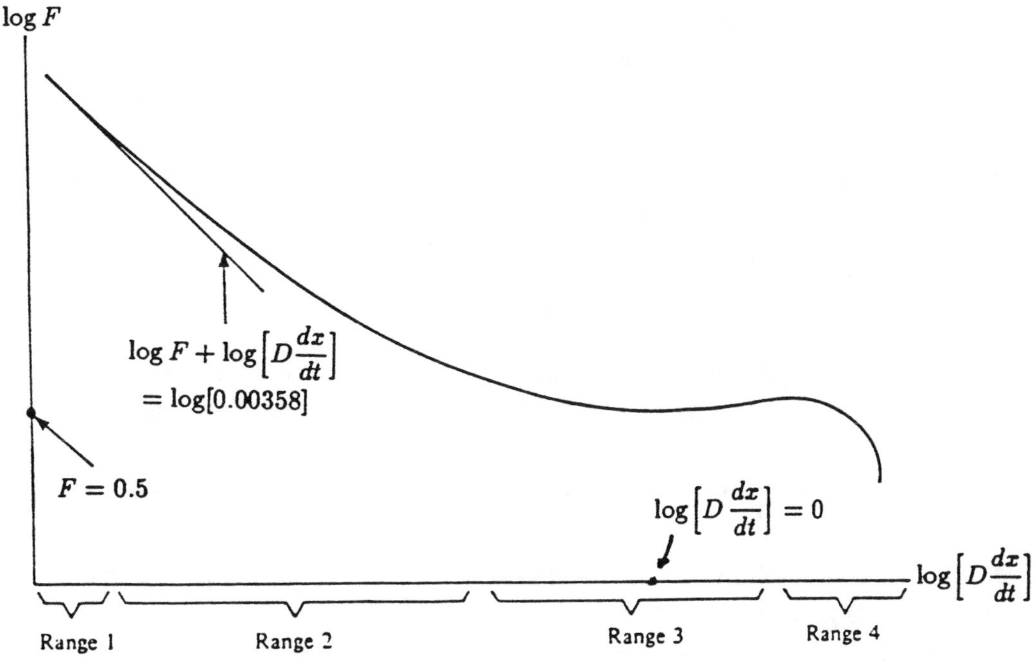

Figure 2

In studying the general air-drag model, it is convenient to distinguish four separate ranges of values for $D\,dx/dt$.

Range 1 (small values of $D\,dx/dt$). Here the plot of $\log F$ versus $\log(D\,dx/dt)$ is approximately linear, with equation

$$\log F + \log\left(D\frac{dx}{dt}\right) = \log 0.00358.$$

From the laws of logarithms, we obtain the equivalent equation

$$F = \frac{0.00358}{D\,dx/dt}.$$

Substituting this into (8) gives (5). Theory and experiment both show that, when $D \leq 0.00025$ feet, the raindrop spends all its time in range 1, so that (5) applies to its entire fall.

Range 2. In this range there is no convenient equation for the function F. For this reason, we cannot simplify (8) to get a convenient differential equation. We could find an inconvenient one, but for our present purposes there would be little learned in doing this.

Range 3. Here $\log F$ is approximately constant, which means F is approximately constant, with a value about 0.5. Substituting this in (8) gives (6). Theory and experiment show that, when $D \geq 0.004$ feet, the raindrop spends nearly all its time in range 3, so that (6) can be assumed to apply to its entire fall.

Range 4. Here again we can't obtain a convenient formula for F. However, it can be shown that this range never applies to raindrops. The reason is that a raindrop would have to be very large to fall in range 4 and raindrops never get that large because they split when their diameters reach about 0.02 feet.

The general drag coefficient function in Figure 2 gives terminal velocities for raindrops which agree well with experimentally determined terminal velocities, as given in Table 2.

Exercises

10. a. Suppose a raindrop spends its entire fall in range 1 of Figure 2 and reaches its terminal velocity so soon after starting its fall that we are willing to assume that it spends its entire fall at this velocity. Under these assumptions, what is the formula for $x(t)$, the distance covered after t seconds of fall? (Your formula can involve D as well as t).

    b. Do the same problem under the assumption that the drop spends practically its entire fall in range 3 and at its terminal velocity.

Table 2. Experimentally determined terminal velocities for raindrops of various sizes.

Drop diameter (feet)	Terminal velocity (feet/second)	Drop diameter (feet)	Terminal velocity (feet/second)
0.00033	0.89	0.00852	24.82
0.00066	2.36	0.00918	25.64
0.00098	3.84	0.00984	26.43
0.00131	5.31	0.01049	27.08
0.00164	6.75	0.01115	27.67
0.00196	8.10	0.01180	28.20
0.00230	9.41	0.01246	28.59
0.00262	10.72	0.01311	28.95
0.00295	12.03	0.01377	29.25
0.00328	13.21	0.01443	29.44
0.00393	15.21	0.01508	29.61
0.00459	16.95	0.01574	29.73
0.00525	18.52	0.01639	29.80
0.00590	19.96	0.01705	29.90
0.00656	21.28	0.01770	29.96
0.00721	22.62	0.01836	30.03
0.00787	23.84	0.01902	30.06

Discussion

Our discussion of models 2, 3, and 4 is meant to show how important air drag can be and how far wrong one can go if one uses model 1, which assumes air drag is not present, in cases where it is substantial. However, it would be wrong to think that air drag is all that prevents equation (3) of model 1 from being exactly correct. Even for objects falling in a vacuum, where there is no drag of any sort, equation (3) is false. The reason is that the acceleration which the earth's gravitation causes in a raindrop or other object is not a constant, but varies with the distance between the object and the earth. The model of variable acceleration, due to Isaac Newton, is known as Newton's Law of Gravitation. Raindrops fall from close enough to the surface of the earth so that the error in assuming that gravitational acceleration is constant is less than the uncertainties in the empirical model 4. Hence we may safely ignore this additional complication. On the other hand, Newton's model becomes critical if we wish to analyze the flight of rockets.

Why not build a completely general model, which would take air drag, variable gravitational acceleration and all other factors into account? Then, instead of choosing a model to fit the particular circumstances, we could use the general model for any situation. Such models have been devised, but they are unwieldy because they require a great deal of calculation. Whenever possible, scientists prefer something simple.

Unfortunately, not all scientific problems can be made simple. A case in point is the calculation of the trajectories of artillery shells and rockets. Here one often needs to use both Newton's law and calculation of air drag. Furthermore, models 2, 3, and 4 for air drag are not adequate since they assume a spherical object traveling through air which has uniform density and no wind currents. We need new and more complex equations that take into account the shape of the object, wind patterns, density of the air at various altitudes, and so on. The situation became so complicated that, during and after World War II, a large number of American mathematicians worked on ballistics for the U.S. Army and produced books of firing tables for every combination of atmospheric factors and types of guns and shells. Often a single trajectory would involve about 12 hours of calculation if done by a human being. The automation of this process became the immediate motivation for, and the initial task of, the world's first electronic digital computer.

When we use mathematics to model natural phenomena, it is usually true that a collection of models of varying degrees of inclusiveness and complication is more helpful than a single model which attempts to be all-inclusive. We can answer some questions with simple models, and save difficult calculations for situations where a simple model fails to capture essential aspects of the problem.

References

Binder, R.C. (1973), *Fluid Mechanics*, Prentice-Hall. Technicalities about drag in air and other fluids.

Blanchard, D.C. (1967), *From Raindrops to Volcanoes*, Anchor Books. This delightful elementary book contains no mathematics, but may inspire some mathematical thoughts. Our exercises on raindrops were motivated by material from this book.

Gunn, Ross, and Gilbert D. Kinzer (1949), "The terminal velocity of fall for water droplets in stagnant air," *Journal of Meteorology* 6, p. 243.

Shapiro, Ascher H. (1961), *Shape and Flow*, Anchor Books. The story behind air drag is here in a form requiring little mathematics, but some physics, for complete comprehension.

This module is adapted from Chapter One, Section 3 of Walter H. Meyer, Concepts of Mathematical Modeling, McGraw-Hill, 1984. It appears with the permission of McGraw-Hill, Inc.

Answers to Exercises

1. a) 400 feet b) 2 seconds c) 5 seconds
2. a) 32 ft/sec, 80 ft/sec b) $v = 8\sqrt{x}$
4. (.00005,.024), (.00010,.098), (.00015,.220) and (.00020,.391) lie on a parabola.
5. c)

$$x(t) = x(0) + \frac{gD^4}{C^2}\left[\frac{Ct}{D^2} + e^{-Ct/D^2} - 1\right]$$

6. b) .087 seconds c) .00087 seconds
7. (.005,18.67), (.006,20.45), (.007,22.08), (.008,23.61), (.009,25.05) and (.010,26.40) lie on a sideways parabola.
8. The differential equation is

$$\frac{dv}{dt} = g\,(1 - \frac{k}{gD}v^2).$$

If we substitute $w = \sqrt{k/gD}v$, we get the integration problem

$$\int \frac{dw}{1-w^2} = \int \sqrt{\frac{kg}{D}}\,dt.$$

Use integration by partial fractions on the lefthand integral.

9. 1.38 seconds
10. a) $x(t) = 9.79 \times 10^6 D^2 t$ b) $x(t) = 264\sqrt{D}\,t$

MEASURING VOTING POWER

Author: Philip Straffin, Beloit College, Beloit, WI 53511

Area of application: political science

Calculus needed: average value of a function, integration of polynomials; mathematical induction.

Related mathematics: elementary probability.

The Problem: Amending the Canadian Constitution

In 1971 the Premiers of the ten provinces of Canada met in Victoria, British Columbia, to negotiate an amendment procedure for the Canadian constitution. The history is interesting. The constitution of Canada was contained in the British North America Act of 1867, by which Britain granted independence to Canada. However, no procedure had been given by which Canadians could amend their constitution, other than by petitioning the British Parliament to enact the amendment. This was a strange situation, and "patriation of the constitution"—bringing the constitution under Canadian control—had been a patriotic issue in Canada since the 1920's. In order to patriate the constitution, the Canadian provinces had to agree on an amending procedure, and this problem was to be addressed by the Victoria Conference.

The problem is complicated by the diversity of the Canadian provinces, in size as well as in politics and culture. The second column of Table 1 shows that in 1970 the two largest provinces, Ontario and Quebec, contained 64% of the Canadian population. Any scheme which treated all provinces equally (as in the United States, where a constitutional amendment must be approved by 3/4 of the states, with large and small states treated equally) would surely be unfair to residents of these provinces.

The amending procedure proposed by the Victoria Conference recognized provincial disparity. A constitutional amendment would have to be approved by

1. Ontario

and 2. Quebec

and 3. British Columbia and one prairie province, <u>or</u> all three prairie provinces

and 4. at least two of the four Atlantic provinces.

Notice that both Ontario and Quebec have veto power over constitutional amendments. British Columbia also seems to have considerable power. The "prairie provinces" are Alberta, Saskatchewan and Manitoba; the "Atlantic provinces" are New Brunswick, Nova Scotia, Prince Edward Island and Newfoundland.

How fair is this scheme? Does the unequal power of the provinces mirror, at least roughly, their relative populations? To answer questions like these, we need to formalize and quantify the notion of the *power* of a voter in a voting body. In this module we will use elementary probability and calculus to develop such a *power index.* The index we will define has seen wide use in many areas of political science.

	Percentage of Population		Percentage of Power under		
Province	1970	1981	Victoria	Lougheed	1982 Act
Ontario	34.9	35.5	31.5	29.1	14.4
Quebec	28.9	26.5	31.5	29.1	12.9
British Columbia	9.4	11.3	12.5	10.0	10.3
Alberta	7.3	9.2	4.2	10.0	9.1
Saskatchewan	4.8	4.0	4.2	5.9	9.1
Manitoba	4.8	4.2	4.2	5.9	9.1
New Brunswick	3.1	2.9	3.0	2.5	8.7
Nova Scotia	3.8	3.5	3.0	2.5	9.1
Prince Edward Island	0.5	0.5	3.0	2.5	8.7
Newfoundland	2.5	2.3	3.0	2.5	8.7

Table 1. Power under Canadian Constitutional Amendment Schemes.

Voting Bodies

A mathematical model of a voting body strips away all personalities and ideologies, and considers only which groups of voters can pass bills (or constitutional amendments in the above case). Those subsets of voters which can pass bills are called *winning coalitions.*

Perhaps the most common kind of voting situation is one in which each voter casts one vote, and a majority of votes is necessary to pass a bill. In other words, the winning coalitions are exactly those which contain more than half of the voters. However, there are voting bodies in which members cast different numbers of votes. Such a body is called a *weighted voting body*, and is described by the symbol $[q; w_1, w_2, ..., w_n]$. Here there are n voters, the ith voter casts w_i votes, and a quota of q votes is needed to pass a bill. For example, the symbol

$$\begin{matrix} [7; & 4, & 3, & 2, & 1] \\ & A & B & C & D \end{matrix} \tag{1}$$

represents a body in which there are four voters (call them A, B, C, D) casting 4, 3, 2, 1 votes respectively, and 7 votes are necessary to pass a bill.

Weighted voting bodies are fairly common. Most familiar to Americans is the United States Electoral College,

$$\begin{matrix} [270; & 47, & 36, & \ldots, & 3, & 3] \\ & CA & NY & & VT & WY \end{matrix} \tag{2}$$

Other classic examples include the Council of Ministers of the European Community, the World Bank, several United Nations organizations, and many county boards in New York state. See (Lucas, 1983) for a wealth of other examples. A legislature in which there are representatives of several parties, who vote under strict party discipline, can also be thought of as a weighted voting body. In this interpretation, example (1) might represent a legislature of 10 members, with 4 belonging to Party A, 3 to Party B, 2 to Party C, and 1 to Party D, with the requirement that a 2/3 majority (7 votes of 10) is needed to pass a bill.

The Victoria Scheme is not a weighted voting scheme. However, its rules do exactly specify the winning coalitions.

Exercise

1. Write out all the winning coalitions in Example (1).

Measuring Power in Voting Bodies

The naive way to think of the distribution of power in a weighted voting body like that of (1) would be to suppose that power is in strict proportion to the number of votes. Thus, A has 40% of the votes and hence should have 40% of the power. A little reflection should convince you that this is not reasonable. For instance, note that in (1) A has veto power: even if B, C and D all favor a bill, it cannot pass without A's approval. This should lead us to believe that A might well have more than 40% of the voting power in this game. Two even more compelling examples are

$$\begin{matrix}[6; & 7, & 1, & 1, & 1] \\ & A & B & C & D\end{matrix} \tag{3}$$

$$\begin{matrix}[6; & 3, & 3, & 3, & 1] \\ & A & B & C & D\end{matrix} \tag{4}$$

In (3) A has 70% of the votes, but she clearly has all of the power. A is a *dictator*, in the sense that a bill passes if and only if A votes for it. In (4) D has 10% of the votes, but no power. D's vote can never make any difference to the outcome, and D is called a *dummy*. In (3) B, C and D are all dummies.

If voting power in a weighted voting body is not proportional to numbers of votes, how can we define and measure it precisely? We will start by thinking of the voting power of a voter as the *probability that that voter's vote will make a difference* to the outcome of a vote on a bill. In other words, voter i's power will be the probability that a bill will pass if voter i votes for it, but would fail if voter i votes against it.

To calculate this probability, we will need to remember the following properties of probability:

i) The probability p of an event $\mathcal{E}$ happening is a number in [0,1]. If $\mathcal{E}$ can never happen, $p = 0$; if $\mathcal{E}$ is certain to happen, $p = 1$.

ii) If p is the probability of $\mathcal{E}$, then the probability that $\mathcal{E}$ will not happen is $1 - p$.

iii) (Sum law) Suppose that p is the probability of $\mathcal{E}$, and q is the probability of $\mathcal{F}$, and $\mathcal{E}$ and $\mathcal{F}$ are *disjoint* events—if one happens, the other cannot. Then the probability that either $\mathcal{E}$ or $\mathcal{F}$ happens is $p + q$.

iv) (Product law) Suppose that p is the probability of $\mathcal{E}$, and q is the probability of $\mathcal{F}$, and $\mathcal{E}$ and $\mathcal{F}$ are *independent* events—whether $\mathcal{E}$ happens has no effect on whether $\mathcal{F}$ happens. Then the probability that both $\mathcal{E}$ and $\mathcal{F}$ happen is pq.

Our idea of voting power as the probability a voter's vote will make a difference assures, by property i), that a dictator will have voting power 1, a dummy will have voting power 0, and all other voters will have power between 0 and 1. To go farther, we need to make some assumptions about how voters vote. Let us suppose that each voter will vote "yes" on a bill with some probability p $(0 \le p \le 1)$, independently of how other voters vote. We can then use properties ii)–iv) to calculate each voter's voting power, as a function of p. Let's start with a simple example:

$$\begin{matrix} [3; & 2, & 1, & 1] \\ & A & B & C \end{matrix} \tag{5}$$

A's vote will make a difference to the outcome of a vote by this body if either B or C, or both, vote yes. (If both B and C vote no, the bill will fail regardless of how A votes.) If all voters vote yes with probability p, the probability that this will happen is

$$f_A(p) = \underset{\text{B yes, C no}}{p(1-p)} + \underset{\text{B no, C yes}}{(1-p)p} + \underset{\text{B,C yes}}{p^2} = 2p - p^2.$$

Similarly, B's vote will make a difference to the outcome if A votes yes and C votes no. (If A votes no, the bill will fail regardless of how B votes; if A and C both vote yes, the bill will pass regardless of how B votes.) Thus

$$f_B(p) = \underset{\text{A yes, C no}}{p(1-p)} = p - p^2.$$

By symmetry, we also have $f_C(p) = p - p^2$.

For a general voting body, the polynomial $f_i(p)$ is called the *power polynomial* for voter i, and it contains interesting information about the ability of voter i to influence the outcome of a vote. However, it might be more useful to have a single number as our measure of voter i's power. One reasonable way to get such a number would be to take the *average value* of $f_i(p)$ over all values of p between 0 and 1. Of course, this is where calculus enters the picture, since you will remember from calculus

Definition. *The average value of a continuous function f(x) on [a,b] is* $\dfrac{1}{b-a}\displaystyle\int_a^b f(x)\,dx.$

We are led to make the following

Definition. *The voting power of voter i is given by* $\phi_i = \int_0^1 f_i(p)\,dp$.

Use of the Greek letter ϕ to denote this measure of voting power is traditional. Hence in example (5),

$$\phi_A = \int_0^1 f_A(p)\,dp = \int_0^1 (2p - p^2)\,dp = 2/3$$

$$\phi_B = \phi_C = \int_0^1 (p - p^2)\,dp = 1/6.$$

See Figure 1. Notice that the voting power of the three voters in this body adds to 1. This will always happen, though it is not at all obvious from the definition. We will see why after we do some more examples.

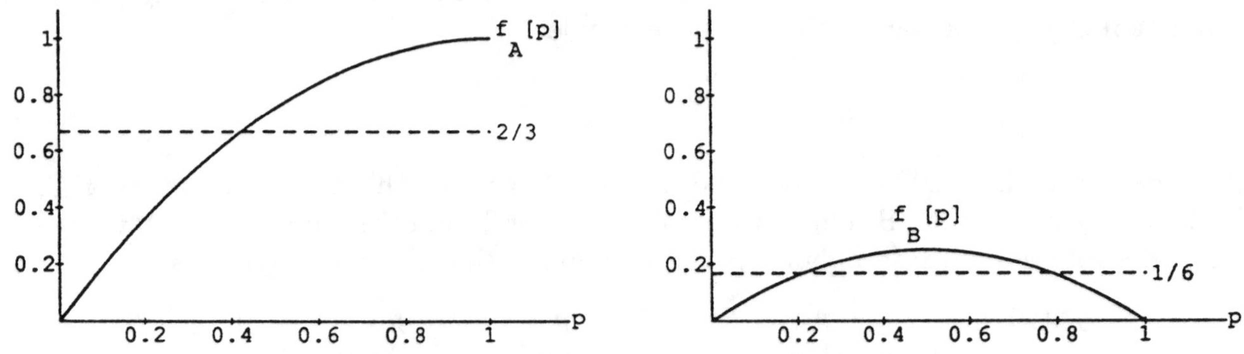

Figure 1. Power polynomials and their average values on [0,1].

For a second example, let's compute the power of the voters in example (1). First, note that A's vote will matter if B votes yes, for then the bill will pass if A votes yes and fail if A votes no, regardless of what C and D do. If B votes no, the bill cannot pass unless both C and D vote yes, and if they do, A's vote will decide the outcome. Hence

$$f_A(p) = \underbrace{p}_{\text{B yes}} + \underbrace{(1-p)\cdot p^2}_{\text{B no, C and D yes}} = p + p^2 - p^3$$

and

$$\phi_A = \int_0^1 (p + p^2 - p^3)dp = 7/12.$$

Similarly, you should check that

$$f_B(p) = p - p^3, \qquad \phi_B = 3/12$$

$$f_C(p) = f_D(p) = p^2 - p^3, \qquad \phi_C = \phi_D = 1/12.$$

Thus A, with 40% of the votes, has 58% of the power. C and D, with different numbers of votes, have the same power. This simply reflects the fact that C and D play symmetric roles in this voting body: C's two votes are no more likely to be decisive to the outcome than D's one vote.

Exercises

2. In the Nassau County (NY) Board of Supervisors in 1964, votes were assigned to representatives of six towns in proportion to the populations of the towns. The result was a weighted voting body

$$[58; 31, 31, 21, 28, 2, 2].$$

Comment on the fairness of this system. You shouldn't need to calculate power polynomials to do this. See (Banzhaf, 1965).

3. Calculate the power polynomials and the power of voters in

$$a.\ \begin{matrix}[3; & 2, & 1, & 1, & 1]\\ & A & B & C & D\end{matrix} \qquad b.\ \begin{matrix}[4; & 3, & 1, & 1, & 1]\\ & A & B & C & D\end{matrix}$$

You can check your answers by remembering that the voters' powers must add to 1.

4. Calculate the power polynomials and the power of voters in

$$a.\ \begin{matrix}[51; & 31, & 27, & 22, & 20]\\ & A & B & C & D\end{matrix} \qquad b.\ \begin{matrix}[67; & 55, & 17, & 15, & 13]\\ & A & B & C & D\end{matrix}$$

$$c.\ \begin{matrix}[67; & 45, & 25, & 15, & 15]\\ & A & B & C & D\end{matrix}.$$

The answers you get should look familiar. What's going on here? Formulate a definition for when two weighted voting bodies should be considered "equivalent."

5. Can changing the quota of a voting body change the power of the members? Calculate the power of voters in

$$\begin{matrix}[6; & 4, & 3, & 2, & 1]\\ & A & B & C & D\end{matrix}$$

and compare to example (1). What about [9;4,3,2,1] and [10;4,3,2,1]? (For these two bodies you shouldn't need to do any calculating.)

Power in the Victoria Amendment Scheme

To see how fairly the Victoria scheme would distribute power among the Canadian provinces, we will compute the power of Ontario (O)—Quebec (Q) would be the same—British Columbia (B), a prairie province (P), and an Atlantic province (A). First, we note that Ontario's vote will make a difference to the outcome if

i) Q votes yes
and ii) B votes yes and at least one P votes yes, or B votes no and all 3 P's vote yes
and iii) at least two A's vote yes.

Translated into probabilities, this gives us the power polynomial for Ontario:

$$f_O(p) = \underbrace{p}_{\text{Q yes and}} \cdot \underbrace{[p\,(1-(1-p)^3) \;+\; (1-p)\,p^3]}_{\text{[B yes, not 3 P's no or B no, 3 P's yes] and}} \cdot \underbrace{[6p^2(1-p)^2 + 4p^3(1-p) + p^4]}_{\text{[at least two A's yes]}}$$

$$= p\,[3p^2 - 2p^3]\,[6p^2 - 8p^3 + 3p^4] = 18p^5 - 36p^6 + 25p^7 - 6p^8.$$

This is harder than the derivations in the last section, but it is still based on the basic probability laws. Perhaps least transparent are the coefficients "6" and "4" in the Atlantic provinces term. The probability that exactly two A's will vote yes (and the other two no) is $p^2(1-p)^2$ times the number of ways we can choose which two of the four A's will vote yes. This number is the *binomial coefficient* $\binom{4}{2} = \frac{4\cdot 3}{1\cdot 2} = 6$ (see Exercise 6). Similarly the coefficient "4" is the binomial coefficient $\binom{4}{3}$, the number of ways of choosing three of the four A's to vote yes.

Hence

$$\phi_O = \int_0^1 f_O(p)\,dp = \frac{18p^6}{6} - \frac{36p^7}{7} + \frac{25p^8}{8} - \frac{6p^9}{9}\Big|_0^1 = 3 - \frac{36}{7} + \frac{25}{8} - \frac{2}{3} = \frac{53}{168} = .3155\,.$$

The calculations for B and P are given below. Notice that $f_P(p)$ is the probability that one given prairie province, say Manitoba, will make a difference. I'd recommend examining the polynomials closely, being sure you see where each term comes from. In Exercise 7, you are asked do the calculations for an Atlantic province.

$$f_B(p) = \underbrace{p^2}_{\text{O,Q yes and}} \cdot \underbrace{[3p(1-p)^2 + 3p^2(1-p)]}_{\text{[1 or 2 P's yes] and}} \cdot \underbrace{[6p^2 - 8p^3 + 3p^4]}_{\text{[at least 2 A's yes (see above)]}}$$

$$= 3\,(6p^5 - 14p^6 + 11p^7 - 3p^8)$$

$$\phi_B = \int_0^1 f_B(p)\,dp = 3\,\left(\frac{6}{6} - \frac{14}{7} + \frac{11}{8} - \frac{3}{9}\right) = \frac{1}{8} = .1250$$

$$f_P(p) = \quad p^2 \quad \cdot \quad [p\,(1-p)^2 \quad + \quad (1-p)\,p^2] \quad \cdot \quad [6p^2 - 8p^3 + 3p^4]$$

O,Q yes and [B yes, other P's no or B no, other P's yes] and [at least 2 A's yes]

$$= 6p^5 - 14p^6 + 11p^7 - 3p^8 \qquad \text{(notice this is 1/3 of } f_B(p))$$

$$\phi_P = \frac{1}{3} \cdot \frac{1}{8} = \frac{1}{24} = .0417$$

The results of our calculations are shown in column 4 of Table 1. Notice how remarkably well the distribution of power under the Victoria scheme matches the relative population of the provinces in 1970. The most serious discrepancy was the underrepresentation of Alberta. During the 1970's Alberta was the fastest growing province in Canada, and the discrepancy would be even more serious by the 1981 census. One method of addressing the problem would be to change condition 3. of the amendment scheme to

3′. British Columbia and Alberta, or one of these together with the other two prairie provinces.

A scheme like this was proposed by Premier Peter Lougheed of Alberta. The power distribution under the Lougheed scheme is shown in column 5 of Table 1. It is an improvement.

Alas, I have to report that neither the Victoria nor the Lougheed scheme was adopted by Canada. Because of considerations unrelated to the constitutional amendment scheme, patriation of the constitution was not accomplished until 1982. The amendment scheme included in the 1982 Constitution Act required approval by at least 7 provinces which together contain at least 50% of the Canadian population. Notice that this scheme does not give veto power to either Ontario or Quebec. Its distribution of power is shown in column 6 of Table 1. It treats the provinces close to equally, and hence seems quite unfair to the large provinces. Sometimes politics just isn't rational.

Exercises

6. Write out the ways to choose two provinces from NB, NS, PEI, NFD. Explain why the number of ways is given by $\frac{4 \cdot 3}{1 \cdot 2}$.

7. Show that $f_A(p) = 3\,(3p^5 - 8p^6 + 7p^7 - 2p^8)$ and $\phi_A = .0298$. (Notice that the derivation in the text of $f_O(p)$ shows how to handle the western provinces, and that for an A's vote to matter, exactly one of the other three A's must vote yes.)

8. Derive the power polynomial and check the power calculation for one province of your choice, under the Lougheed scheme.

The Combinatorial Shapley-Shubik Power Index

The voting power index calculated as in the last section has an interesting history. It was first defined by Lloyd Shapley and Martin Shubik in purely combinatorial terms:

> There is a group of individuals all willing to vote for some bill. They vote in order. As soon as a [quota] has voted for it, it is declared passed, and the member who voted last is given credit for having passed it. Let us choose the voting order randomly. Then we may compute the frequency with which an individual ... is pivotal. This latter number serves to give us our index. (Shapley and Shubik, 1954)

The calculation would be as follows for example (6). Write down all 3! orders in which the voters could vote for a bill, and in each order underline the voter who is ***pivotal*** (i.e. whose vote passes the bill):

$$A\underline{B}C \quad A\underline{C}B \quad B\underline{A}C \quad BC\underline{A} \quad C\underline{A}B \quad CB\underline{A}$$

For example, in ordering BCA, A is pivotal since B and C together have just two votes, and A brings in the third vote necessary to pass the bill. To calculate the power of a voter, count the number of times that voter is pivotal (underlined) and divide by the number of orderings. Let us temporarily denote this power index for voter i by SS_i. Thus $SS_A = 4/6$, and $SS_B = SS_C = 1/6$. Notice that these are exactly the same numbers that we obtained by the completely different approach of integrating power polynomials. In fact, we will show below that $SS_i = \phi_i$.

The equivalence of the Shapley-Shubik approach and the power polynomial approach was shown in (Straffin, 1977), which introduced the power polynomial. It was important because the Shapley-Shubik power index, although widely used, was thought to depend upon the voters voting in order, or at least joining a coalition supporting a bill in order. The power polynomial approach shows that the index can be defined with no reference to orderings. On the other hand, it is obvious from the Shapley-Shubik definition that the powers of the voters add to 1, which is not obvious from the power polynomial definition. The two approaches complement each other. The proof of equivalence is given in the next section.

Exercise

9. Use the Shapley-Shubik definition to calculate the Shapley-Shubik power indices for the voting bodies in Exercise 3.

Proof of Equivalence (Optional)

Theorem. *For a voter i in any voting game, $\phi_i = SS_i$.*

Start of Proof: We will start by deriving a general expression for $f_i(p)$.

Suppose that in a vote for some bill, S is the set of yes voters, and assume that i votes yes, so that i is in S. Now i's vote will matter exactly if S is a winning coalition, but $S-i$ is not (so that the bill passes when i votes for it, but would fail if i switched his vote). If this happens, we say i is a *swing vote* in S, or (for short) "i swings in S."

Now let n be the total number of voters, s be the size of S, and suppose that all voters vote yes with probability p. The probability that the set of voters who vote yes will be exactly S is $p^{s-1}(1-p)^{n-s}$, since the $s-1$ other voters in S would all have to vote yes (we already know that i is voting yes), and the $n-s$ voters not in S would all have to vote no. We thus get that

$$f_i(p) = \sum_{i \text{ swings in } S} p^{s-1}(1-p)^{n-s}.$$

The notation means that the sum is to be taken over all coalitions S in which i is a swing vote. Averaging over [0,1] gives

$$\phi_i = \int_0^1 \sum_{i \text{ swings in } S} p^{s-1}(1-p)^{n-s}\,dp = \sum_{i \text{ swings in } S} \int_0^1 p^{s-1}(1-p)^{n-s}\,dp.$$

To proceed further we need to evaluate the integrals. Definite integrals of this type have a nice form related to a function from higher analysis known as the "Beta function":

Lemma. $B(a,b) = \int_0^1 x^a(1-x)^b\,dx = a!\,b!/(a+b+1)!$

(Here the notation $a!$ means "a factorial": $a! = a(a-1)(a-2)\ldots 3\cdot 2\cdot 1$.)

Proof of Lemma: Our proof will be by mathematical induction on b.

If $b=0$, direct integration shows that $B(a,0) = 1/(a+1)$, so the lemma is true in this case. (Recall that $0! = 1$.)

Now assume that the lemma is true whenever the second index is b. We will show that it is true when the second index is $b+1$. This will complete the induction proof.

$$\begin{aligned}
B(a,b+1) &= \int_0^1 x^a(1-x)^{b+1}\,dx = \int_0^1 x^a(1-x)^b[1-x]\,dx \\
&= \int_0^1 x^a(1-x)^b\,dx - \int_0^1 x^{a+1}(1-x)^b\,dx = B(a,b) - B(a+1,b) \\
&= \frac{a!\,b!}{(a+b+1)!} - \frac{(a+1)!\,b!}{(a+b+2)!} \quad \text{by the induction assumption} \\
&= \frac{a!\,b!}{(a+b+2)!}[(a+b+2)-(a+1)] = \frac{a!\,(b+1)!}{(a+b+2)!}.
\end{aligned}$$

Since this is what the lemma says should happen when the second index is $b+1$, the proof of the lemma is complete.

Continuation of Proof of Theorem: Using the result of the lemma, we have

$$\phi_i = \sum_{i \text{ swings in } S} \frac{(s-1)!\,(n-s)!}{n!} = \frac{1}{n!} \cdot \sum_{i \text{ swings in } S} (s-1)!\,(n-s)!$$

The $n!$ in the denominator is just the total number of orderings of n voters. I claim that the sum is exactly the number of orderings in which voter i is pivotal. Indeed, voter i will be pivotal in an ordering exactly when i and the voters who precede him form a swing coalition for i, so that i brings in the winning vote. The number of orderings which give rise to a particular swing coalition S is exactly $(s-1)!$ (the number of ways to arrange the $s-1$ voters who precede i) times $(n-s)!$ (the number of ways to arrange the $n-s$ voters who follow i).

Hence we have shown that $\phi_i = SS_i$.

Exercise

10. Evaluate $\int_0^1 x^4(1-x)^5\,dx$.

Further Reading

For many examples of voting bodies and powerful methods for calculating the Shapley-Shubik power index, see (Lucas, 1983) and (Lambert, 1988). For discussions of the Canadian constitutional amendment schemes, see (Straffin, 1977) and (Kilgour and Levesque, 1984). A power analysis of the United States Electoral College is given in (Owen, 1975). (Brams, 1976) gives interesting examples of paradoxes of voting power. (Straffin, 1977, 1983) give comparative discussions of the Shapley-Shubik index and other proposed indices of voting power. (Straffin, 1983) also has a table of the power polynomials of all voting games with four or fewer voters.

References

Brams, Steven (1976), *Paradoxes in Politics*, Free Press.

Kilgour, D.M. and T.J. Levesque (1984), "The Canadian constitutional amending formula: bargaining in the past and future," *Public Choice* 44: 457-480.

Lambert, J.P. (1988), "Voting games, power indices, and Presidential elections," *UMAP Journal* 9: 213-267.

Lucas, William (1983), "Measuring power in weighted voting systems," in Brams, Lucas and Straffin, eds., *Political and Related Models*, Springer-Verlag: 183-238.

Owen, Guillermo (1975), "Evaluation of a Presidential election game," *American Political Science Review* 69: 947-953.

Shapley, Lloyd and Martin Shubik (1954), "A method for evaluating the distribution of power in a committee system," *American Political Science Review* 48: 787-792.

Straffin, Philip (1977), "Homogeneity, independence and power indices," *Public Choice* 30: 107-118.

Straffin, Philip (1979), "Using integrals to evaluate voting power," *Two-Year College Mathematics Journal* 10: 179-181.

Straffin, Philip (1983), "Power indices in politics," in Brams, Lucas and Straffin, eds., *Political and Related Models*, Springer-Verlag: 256-321.

Answers to Exercises

1. AB, ABC, ABD, ACD, ABCD.
2. The vote on any issue is decided by the largest three voters. The representative with 21 votes and the two representatives with 2 are all dummies: the towns they represent have no power at all. In 1965, John Banzhaf, who lived in one of these towns, sued Nassau County and won, establishing the legal precedent that a voting scheme must apportion voting *power* fairly.
3. a.

$$f_A(p) = \underbrace{3p(1-p)^2 + 3p^2(1-p)}_{\text{1 or 2 of B,C,D yes}} = 3p - 3p^2 \qquad \phi_A = 1/2$$

$$f_B(p) = \underbrace{p(1-p)^2}_{\text{A yes; C,D no}} \text{ or } + \underbrace{(1-p)p^2}_{\text{A no; C,D yes}} = p - p^2 \qquad \phi_B = 1/6$$

   b.

$$f_A(p) = \underbrace{1-(1-p)^3}_{\text{not all of B,C,D no}} = 3p - 3p^2 + p^3 \qquad \phi_A = 3/4$$

$$f_B(p) = \underbrace{p(1-p)^2}_{\text{A yes; C,D no}} = p - 2p^2 + p^3 \qquad \phi_B = 1/12$$

4. These bodies have the same winning coalitions, hence the same power indices, as the bodies in Exercises 3a, 3b, and 1 respectively. Two voting bodies with the same players are *equivalent* if they have exactly the same winning coalitions.
5. [6;4,3,2,1] has power indices 5/12, 3/12, 3/12, 1/12. [9;4,3,2,1] has power indices 1/3, 1/3, 1/3, 0. [10;4,3,2,1] has power indices 1/4, 1/4, 1/4, 1/4. Yes, changing the quota can seriously change the distribution of power.
6. NB,NS; NB,PEI; NB,NFD; NS,PEI; NS,NFD; PEI,NFD. There are 4 ways to choose the first province, then three ways to choose the second. However, this counts each pair twice (e.g. NB,NS and NS,NB), so we need to devide by two to get the number of pairs.
7.

$$f_A(p) = \underbrace{p^2}_{\text{O,Q yes.}} \cdot \underbrace{[3p^2 - 2p^3]}_{\text{See the derivation of } f_O(p).} \cdot \underbrace{3p(1-p)^2}_{\text{Exactly one other A yes.}}$$

8. Call British Columbia and Alberta "mountain provinces" (M). Then

$$f_O(p) = p\,[p^2 + 2p(1-p)p^2][6p^2 - 8p^3 + 3p^4] = 6p^5 + 4p^6 - 25p^7 + 22p^8 - 6p^9$$
$$f_M(p) = p^2[p(1-p^2) + (1-p)p^2][6p^2 - 8p^3 + 3p^4] = 6p^5 - 2p^6 - 17p^7 + 19p^8 - 6p^9$$
$$f_P(p) = p^2[2p(1-p)p][6p^2 - 8p^3 + 3p^4] = 2\,(6p^6 - 14p^7 + 11p^8 - 3p^9)$$
$$f_A(p) = p^2[p^2 + 2p(1-p)p^2][3p(1-p)^2] = 3\,(p^5 - 5p^7 + 6p^8 - 2p^9)$$

9. a. You could write out 24 orderings and underline the pivot in each one. An easier way is to note that A will pivot when she votes 2nd or 3rd, hence in 12 orderings. By symmetry, the other 12 orderings will have pivots split equally among B, C and D.

   b. Here A pivots when she votes 2nd, 3rd or 4th, hence for 18 orderings, and again the other 6 are divided equally.

10. $4!5!/10! = 1/1260$.

HOW TO TUNE A RADIO

Author: Clark Benson, University of Arizona and the National Security Agency.

Area of application: signal analysis.

Calculus needed: differentiation and integration of sines and cosines.

Remarks on using the module: It is desirable to have access to a computer with a simple plotting routine. Exercises requiring a plotter are marked with an asterisk (*).

The Problem: Tuning A Radio

If you want to listen to WBBM, a Chicago radio station, you tune your radio to "780 on your AM dial." Somehow, that picks out WBBM's signal from the signals of all other radio stations in the Chicago area. The selection process is based on a mathematical problem which can be formulated in terms of functions and solved by calculus.

The signal produced by a radio station can be represented by a function $f(t)$. Here t represents time and $f(t)$ is some kind of time-varying voltage. We will not concern ourselves with the physics—it is enough to know that $f(t)$ can be measured and plotted. Now suppose that we have m stations $S_1, S_2, \ldots, S_m$. Each S_i produces a signal $f_i(t)$. By the time the signals enter our radio, they have been superimposed and the radio receives $f_1(t) + f_2(t) + \ldots + f_m(t)$. We only want to receive one of the signals, say $f_1(t)$. How do we get rid of $f_2(t), f_3(t), \ldots, f_m(t)$?

For general functions, if

$$f(t) = f_1(t) + f_2(t),$$

it is impossible to recover $f_1(t)$ and $f_2(t)$, knowing only $f(t)$. After all, we cannot even solve this kind of problem for numbers: if $x + y = 10$, what are x and y? The fact that we can solve the problem for radio signals is based on the fact that radio signal functions involve sines and cosines in a special way, and the answer involves some very interesting properties of sines and cosines. You will see that a lot goes on when you "turn the dial"! Moreover, in this module you will be introduced to an amazing discovery of Joseph Fourier, made in about 1807—namely that most functions can be written in terms of sines and cosines.

A Partial Solution

We begin with the fact that $\int_0^\pi \cos nt\,dt = 0$ for $n \neq 0$.

Exercises

1. Compute $\int \cos nt\,dt$ and show that if $n \neq 0$ then $\int_0^\pi \cos nt\,dt = 0$.
2. Graph $\cos nt$ and $\sin nt$ for $n = 1,2$, and 3. Can you tell that $\int_0^\pi \cos nt\,dt = 0$ directly from the graph? How? Which of these curves oscillate most rapidly?

We next show that

$$\int_0^\pi \sin nt \sin mt\,dt = 0 \quad \text{and} \quad \int_0^\pi \cos nt \cos mt\,dt = 0 \quad \text{for} \quad n \neq \pm m. \tag{1}$$

You may not have evaluated integrals like these in your class, but they are easy to do if the right trigonometric identities are used. Recall that

$$\begin{aligned} \cos(A-B) &= \cos A \cos B + \sin A \sin B \\ \cos(A+B) &= \cos A \cos B - \sin A \sin B. \end{aligned} \tag{2}$$

Subtracting the second equation from the first gives

$$\cos(A-B) - \cos(A+B) = 2 \sin A \sin B$$

or

$$\sin A \sin B = \frac{1}{2}\left[\cos(A-B) - \cos(A+B)\right]. \tag{3}$$

Thus,

$$\int_0^\pi \sin nt \sin mt\,dt = \frac{1}{2}\int_0^\pi \left[\cos(n-m)t - \cos(n+m)t\right] dt = 0$$

if $n \neq \pm m$ by Exercise 1. Similarly, if we add the equations in (2) and proceed as above, we see that $\int_0^\pi \cos mt \cos nt\,dt = 0$ for $n \neq \pm m$.

Exercise

3. Use the identities $\sin^2 A = \frac{1}{2}(1 - \cos 2A)$ and $\cos^2 A = \frac{1}{2}(1 + \cos 2A)$ to show that

$$\int_0^\pi \sin^2 nt\,dt = \pi/2 \quad \text{and} \quad \int_0^\pi \cos^2 nt\,dt = \pi/2, \quad \text{for} \quad n = 1, 2, \ldots$$

Getting back to our radio problem, we must realize that stations cannot send out any signal that they wish. A station's signal must be designed as follows: every station is assigned a function $\sin nt$. This determines the station's *frequency*, which is what WBBM's "780" on the dial refers to. The station can then send out a signal of the form $g(t)\sin nt$. In practice n is quite large (near a million) and thus $\sin nt$ oscillates very rapidly. The function $g(t)$, which represents the talk, music, etc. that the station is sending varies much more slowly than $\sin nt$. (Actually, cosines as well as sines are used and n need not be an integer, but to keep things simple we will stick with sine curves and integral n.) We will illustrate the principles involved by taking n around 20.

Exercise

*4. Plot $t\sin 20t$, $t^2\sin 20t$, and $\sin t\sin 20t$. Can you find the functions t, t^2, and $\sin t$ hidden in the graphs? Explain.

Exercise 4 shows that if $g(t)$ is slowly varying in comparison to $\sin nt$, then the graph of $g(t)\sin nt$ looks like $\sin nt$ except that the amplitude varies depending on $g(t)$, forming an "envelope" that looks like $\pm g(t)$. This kind of signal is called an *amplitude modulated* signal, or *AM signal.* The information $g(t)$ is "carried" by the rapidly oscillating function $\sin nt$. Thus $\sin nt$ is called the *carrier signal.*

Now let's consider an example of the radio problem. For the "slowly varying" signals $g(t)$ we will use combinations of $\sin t$ and $\sin 2t$. Suppose that station S_1 wants to send $3\sin t - 2\sin 2t$, station S_2 wants to send $5\sin t + 6\sin 2t$, and station S_3 wants to send $4\sin t + 7\sin 2t$. Suppose further that the carrier signals of S_1, S_2, and S_3 are $\sin 8t$, $\sin 16t$, and $\sin 24t$, respectively. Thus, S_1 sends $(3\sin t - 2\sin 2t)\sin 8t$, S_2 sends $(5\sin t + 6\sin 2t)\sin 16t$, and S_3 sends $(4\sin t + 7\sin 2t)\sin 24t$. These signals are superimposed on one another and the signal reaching our radio is

$$f(t) = (3\sin t - 2\sin 2t)\sin 8t + (5\sin t + 6\sin 2t)\sin 16t + (4\sin t + 7\sin 2t)\sin 24t.$$

See Figure 1 for a graph of the individual signals and also for the composite signal reaching the receiving radio. How can we "tune" to, say, station S_2 and recover the function $5\sin t + 6\sin 2t$ which represents the signal being sent from station S_2?

To solve the problem, we use equation (3) to rewrite $f(t)$ as

$$\begin{aligned} f(t) = {} & \frac{3}{2}[\cos 7t - \cos 9t] - \frac{2}{2}[\cos 6t - \cos 10t] \\ & + \frac{5}{2}[\cos 15t - \cos 17t] + \frac{6}{2}[\cos 14t - \cos 18t] \\ & + \frac{4}{2}[\cos 23t - \cos 25t] + \frac{7}{2}[\cos 22t - \cos 26t]. \end{aligned}$$

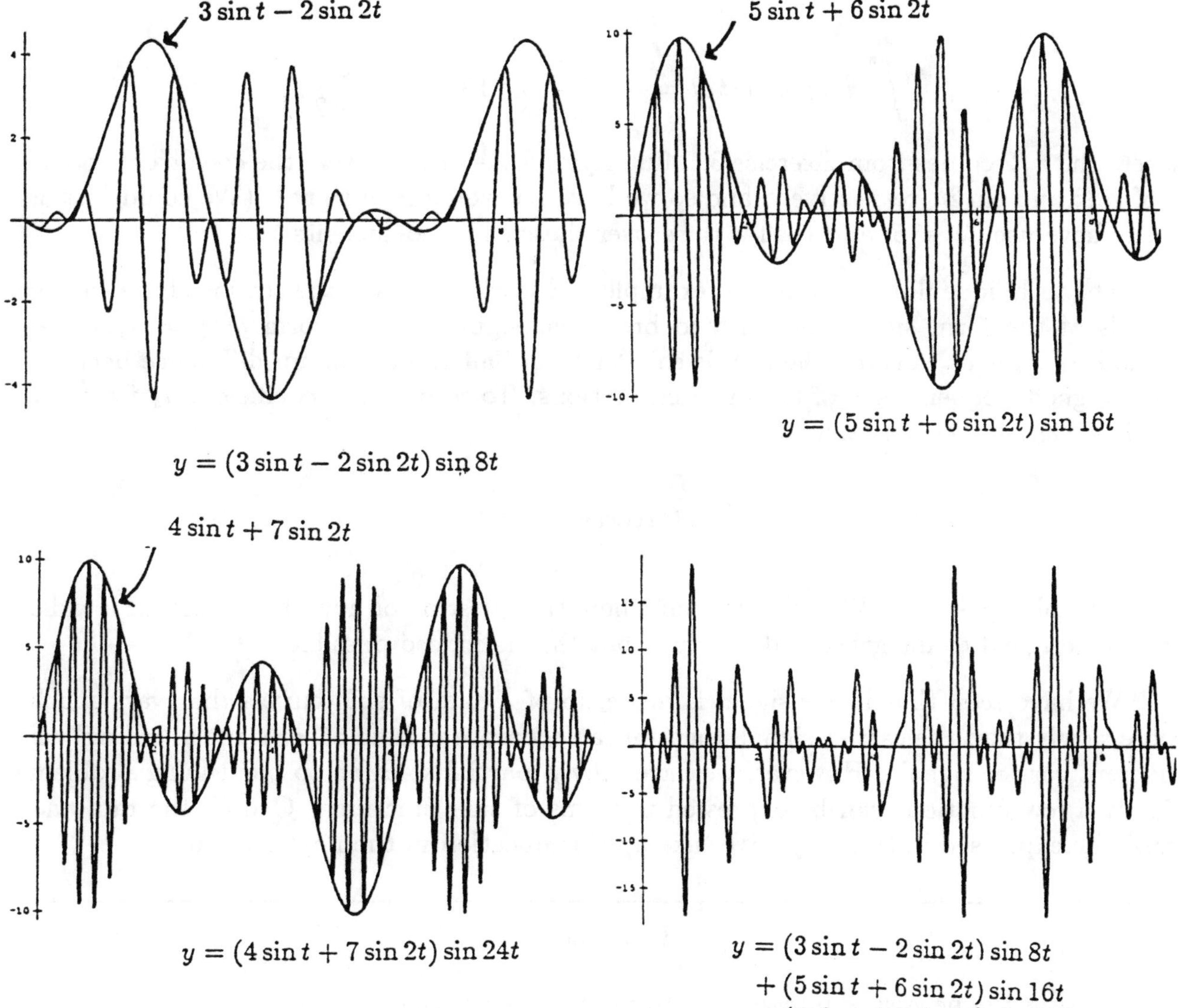

$y = (3\sin t - 2\sin 2t)\sin 8t$

$y = (5\sin t + 6\sin 2t)\sin 16t$

$y = (4\sin t + 7\sin 2t)\sin 24t$

$y = (3\sin t - 2\sin 2t)\sin 8t + (5\sin t + 6\sin 2t)\sin 16t + (4\sin t + 7\sin 2t)\sin 24t$

Figure 1

We then form the integrals

$$\int_0^\pi f(t)\cos 17t\,dt \quad \text{and} \quad \int_0^\pi f(t)\cos 18t\,dt.$$

All but one of the terms in each of the integrals vanish by equation (1). Thus, we have

$$\int_0^\pi f(t)\cos 17t\,dt = -\frac{5}{2}\int_0^\pi \cos^2 17t\,dt = -\frac{5}{2}\frac{\pi}{2}$$

and

$$\int_0^\pi f(t)\cos 18t\,dt = -\frac{6}{2}\int_0^\pi \cos^2 18t\,dt = -\frac{6}{2}\frac{\pi}{2},$$

where the $\pi/2$ comes from Exercise 3. Multiplying by $-4/\pi$ recovers the coefficients 5 and 6 of $\sin t$ and $\sin 2t$, respectively, and so we have recovered S_2's signal. (We could just as easily have used $\cos 15t$ and $\cos 14t$ to recover these same coefficients.)

The technique illustrated in this example will work whenever the stations have carrier signals of the form $\sin nt$ and wish to broadcast signals of the form $g(t) = a_1 \sin t + a_2 \sin 2t + \ldots + a_m \sin mt$, where m is smaller than half the minimum difference between the assigned frequencies n of the different stations. To recover the coefficient a_k from the received signal $f(t)$, we calculate

$$\int_0^\pi f(t)\cos(n+k)t\,dt$$

and multiply by $-4/\pi$. We will not discuss here the question of how the circuits in a radio can be designed to integrate and multiply, but that is indeed possible.

We have seen that if our signal is made up of a sum of sine curves that vary much more slowly than the carrier signal, then we can separate it from other signals "riding" on different carrier signals. However, we have only a partial solution to our tuning problem if only a few functions can be expressed in terms of sine functions. Can all the talk and music be expressed in this way? We take up this question in the next section.

Exercises

5. Perform the necessary integrations to recover S_3's signal.

6. a. Suppose that S_2's signal were $g(t) = 5\sin t + 6\sin 2t + 7\sin 3t$. Write the new form of $f(t)$. What integral would you calculate to recover the coefficient "7" in S_2's signal?

   b. Suppose that S_2's signal were $g(t) = 5\sin t + 6\sin 2t + 7\sin 6t$. Explain why we then could not recover the coefficient "7" by doing an integration.

Fourier Analysis

It may seem that a signal made up of sine curves is very special. However, almost two hundred years ago Joseph Fourier astounded the mathematical world by claiming that he could write "any function" in terms of sines and cosines. Again, for simplicity, we will illustrate his ideas by using only sine curves. Fourier claimed that given $g(t)$ defined on $(0, \pi)$, he could write

$$g(t) = a_1 \sin t + a_2 \sin 2t + a_3 \sin 3t + \ldots \tag{4}$$

Such a sum is called a *Fourier sine series*. There are also *Fourier cosine series* and series involving both cosines and sines (*complete Fourier series*).

Working formally with (4), we multiply both sides by $\sin nt$ and integrate from 0 to π. We obtain

$$\int_0^\pi g(t) \sin nt\, dt = \int_0^\pi a_1 \sin t \sin nt\, dt + \int_0^\pi a_2 \sin 2t \sin nt\, dt + \int_0^\pi a_3 \sin 3t \sin nt\, dt + \ldots.$$

Using equation (1), we see that all the integrals except one vanish on the right hand side and we obtain

$$\int_0^\pi g(t) \sin nt\, dt = a_n \int_0^\pi \sin^2 nt\, dt = a_n \pi/2.$$

Thus the coefficients a_n are given by

$$a_n = \frac{2}{\pi} \int_0^\pi g(t) \sin nt\, dt. \tag{5}$$

As an example let us take $g(t) = 1$ for t between 0 and π. We have

$$a_n = \frac{2}{\pi} \int_0^\pi g(t) \sin nt\, dt = \frac{2}{\pi} \int_0^\pi \sin nt\, dt = \frac{-2}{\pi n} \cos nt \Big|_0^\pi$$

$$= \begin{cases} 0 & \text{if } n \text{ is even} \\ \frac{4}{\pi n} & \text{if } n \text{ is odd.} \end{cases}$$

Thus, we are saying that

$$1 = \frac{4}{\pi} \left[\sin t + \frac{\sin 3t}{3} + \frac{\sin 5t}{5} + \frac{\sin 7t}{7} + \ldots \right] \tag{6}$$

for t between 0 and π. It is interesting to see how a function $g(t)$ is approximated by the first few terms of its Fourier series. Figure 2 shows the graph of the first three terms of the series in (6) and Figure 3 shows the graph of the first six terms of the series.

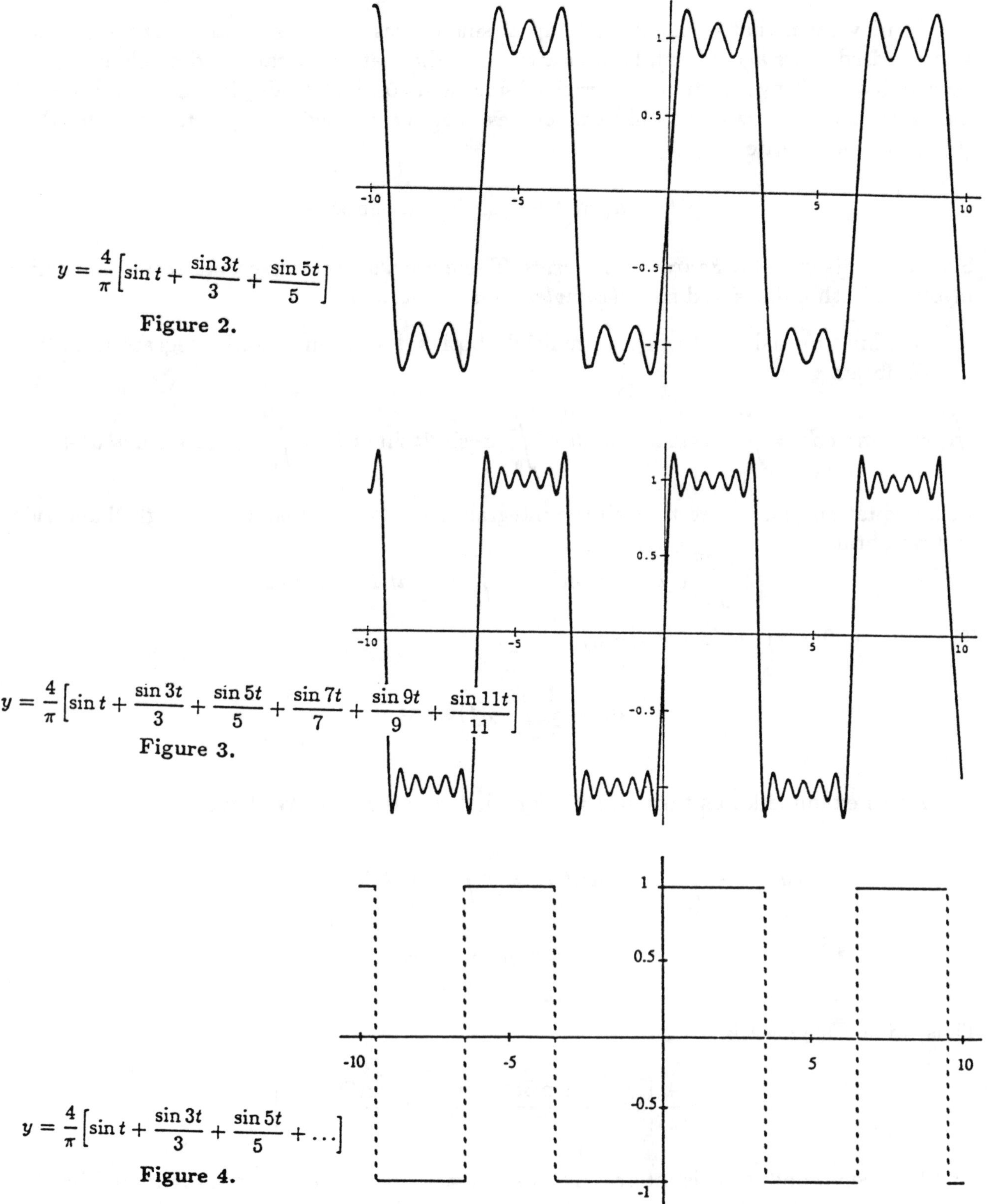

$$y = \frac{4}{\pi}\left[\sin t + \frac{\sin 3t}{3} + \frac{\sin 5t}{5}\right]$$

Figure 2.

$$y = \frac{4}{\pi}\left[\sin t + \frac{\sin 3t}{3} + \frac{\sin 5t}{5} + \frac{\sin 7t}{7} + \frac{\sin 9t}{9} + \frac{\sin 11t}{11}\right]$$

Figure 3.

$$y = \frac{4}{\pi}\left[\sin t + \frac{\sin 3t}{3} + \frac{\sin 5t}{5} + \dots\right]$$

Figure 4.

Note that although we are approximating the function $g(t) = 1$ only on the interval $(0,\pi)$, the sine terms are defined everywhere. Since $\sin(-nt) = -\sin nt$, the sum of the sine terms represents an odd function. Thus on the interval $(-\pi,\pi)$, the sum of the sine terms approximates the function given by

$$\begin{cases} -1 & \text{if} \quad -\pi < t < 0 \\ 0 & \text{if} \quad t = 0 \\ +1 & \text{if} \quad 0 < t < \pi \end{cases}$$

on the interval $(-\pi,\pi)$. Since $\sin nt$ is periodic, we get this pattern over and over. Eventually, as the number of sine terms becomes greater and greater, their sum gets closer and closer to the "square wave" in Figure 4.

Returning to the radio problem, we can see the general solution. A station broadcasts a signal $g(t)$ by broadcasting enough terms in the Fourier series for $g(t)$ to give a good approximation, and our radios sort out those Fourier coefficients. Radio is an application of trigonometric integrals.

Exercises

7. Compute the a_n's for the function $g(t)$ given by

$$g(t) = \begin{cases} 0 & \text{if} \quad 0 < t < \frac{\pi}{2} \\ 1 & \text{if} \quad \frac{\pi}{2} \le t < \pi. \end{cases}$$

*8. Plot the first three terms and the first six terms of the Fourier sine series approximation to the $g(t)$ of Exercise 7.

9. Substitute $t = \pi/2$ in equation (6) and solve for π, to obtain a famous infinite series formula for π derived by Leibniz in 1673.

10. If you know integration by parts, use it to compute the a_n's for the function

$$h(t) = t, \quad 0 \le t < \pi.$$

*11. Plot the first three terms and the first six terms of the Fourier sine series approximation to $h(t)$ of Exercise 10.

Historical Comment

Fourier developed his theory of writing arbitrary functions as infinite sums of sines and cosines as a technique for solving the equations which describe how heat flows. His theory had nothing to do with radio, which wasn't invented until the end of the nineteenth century.

Hence it might seem remarkable that Fourier's analysis turned out to be exactly what was needed to sort out radio signals, except that this phenomenon has been extremely common in the history of science and mathematics. It seems to be in the nature of mathematics that powerful techniques developed for one purpose will solve other problems which haven't yet arisen.

Fourier's work was also a stimulus to pure mathematics. What does it mean to sum an infinite number of sines and cosines? Exactly which functions can be represented by a Fourier series? Does a Fourier series represent its function $g(t)$ everywhere, or only at some places? (The examples and exercises above show that we should expect trouble, for example, at places where $g(t)$ is discontinuous.) The first of these questions will be clarified when you study the theory of infinite series. The others led to many of the most important mathematical developments of the nineteenth century.

References

Arfken, George (1985), *Mathematical Methods for Physicists*, Academic Press. Chapter 14 introduces Fourier series.

Bell, E.T. (1937), *Men of Mathematics*, Simon and Schuster. Chapter 12 contains a rousing account of Fourier's life and his relatioship with Napoleon.

Carr, Joseph (1980), *The Complete Handbook of Radio Transmitters* and *The Complete Handbook of Radio Receivers*, TAB.

Davis, Philip and Reuben Hersh (1980), *The Mathematical Experience*, Birkhauser. Pages 255-270 discuss the history of Fourier analysis and its effect on mathematics.

Erst, Stephen (1984), *Receiving Systems Design*, Artech House. This discusses not just AM signals, but also frequency modulation (FM) and phase modulation.

Gray, Paul and Campbell Searle (1969), *Electronic Principles: Physics, Models and Circuits*, Wiley. Information about simple electronic circuits which do integration and multiplication.

Haberman, Richard (1987), *Elementary Applied Partial Differential Equations, with Fourier Series and Boundary Value Problems*, 2nd edition, Prentice-Hall. Chapter 3 introduces Fourier series.

Terman, Fredrick (1947), *Radio Engineering*, McGraw-Hill. Chapter 1.

Answers to Exercises

4.

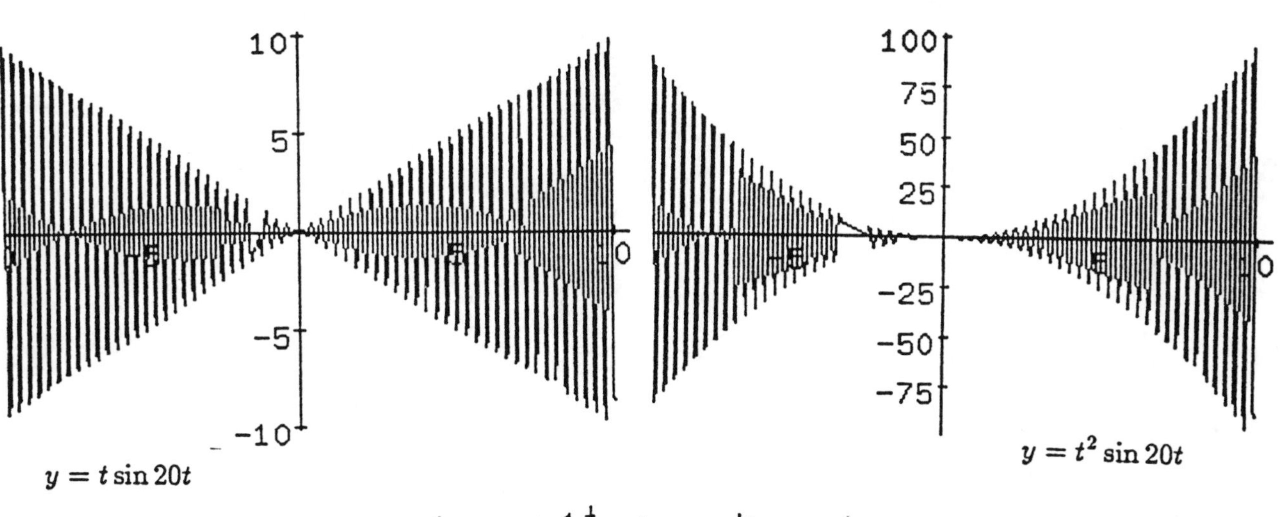

$y = t \sin 20t$ $y = t^2 \sin 20t$

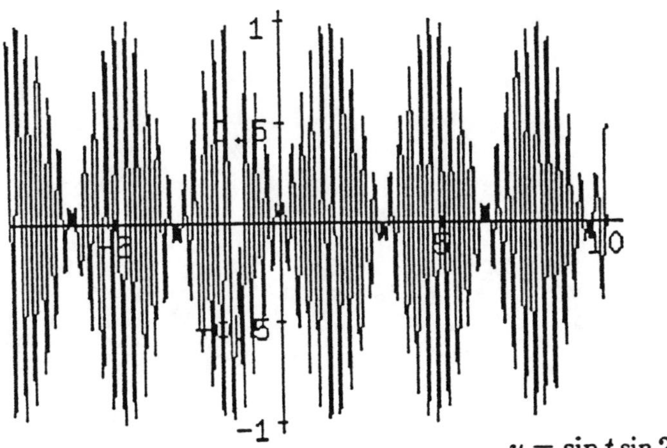

$y = \sin t \sin 20t$

6. a. $\int_0^\pi f(t) \cos 19t \; dt$ or $\int_0^\pi f(t) \cos 13t \; dt$.

b. The signal would overlap the signals from stations 1 and 3.

7.

$$a_n = \frac{2}{\pi n} \text{ if } n \text{ is odd}, \quad \frac{-4}{\pi n} \text{ if } n \text{ is even}.$$

$$f(t) = \frac{2}{\pi}[\sin t - \sin 2t + \frac{1}{3}\sin 3t - \frac{1}{2}\sin 4t + \frac{1}{5}\sin 5t - \frac{1}{3}\sin 6t + \ldots].$$

10.

$$\begin{aligned} a_n &= \frac{2}{\pi}\left[\frac{-t}{n}\cos nt\Big|_0^\pi + \frac{1}{n}\int_0^\pi \cos nt\ dt\right] \\ &= \frac{2}{n} \text{ if } n \text{ is odd,} \quad \frac{-2}{n} \text{ if } n \text{ is even.} \end{aligned}$$

$$g(t) = 2[\sin t - \frac{1}{2}\sin 2t + \frac{1}{3}\sin 3t - \frac{1}{4}\sin 4t + \ldots].$$

VOLUMES AND HYPERVOLUMES

Author: Philip Straffin, Beloit College, Beloit, WI 53511

Area of application: geometry

Calculus needed: Riemann sums, definite integrals, integration by substitution. Multiple integrals are not used.

Suggestions for use: This could be used in place of standard sections on Solids of Revolution. I have tried to focus on the geometric problem of volume in some of its interesting incarnations.

The Problem: Computing Volumes

Any student of calculus is familiar with the use of integrals to compute areas of planar regions. Indeed, we usually picture a definite integral $\int_a^b f(x)dx$ as the area under the curve $y = f(x)$ between $x = a$ and $x = b$, at least as long as $f(x) \geq 0$ on $[a, b]$. However, computing areas has been less important in history, and is less important today, than the harder problem of computing *volumes* of solid figures. After all, we live in a three dimensional space, keep our belongings in three dimensional containers, and build three dimensional structures. This module explores the power of integrals in calculating volumes, via a fundamental principle we will call the *slicing principle.*

We will start by considering simple geometrical shapes whose volumes are given by formulas often taught in elementary school. Those formulas were not as easy to derive and prove as they are to state. Each was a triumph of mathematical intuition and argument when it was first discovered. Integral calculus and the slicing principle give a straightforward way of deriving all of them, and also enable us to calculate more complicated volumes.

Having seen the power of integrals in three dimensions, we'll then ask what they can do in higher dimensions. What is the hypervolume of a hypersphere in four dimensional space? Five dimensional space? The slicing principle supplies the answers.

The Slicing Principle

The easiest kind of volume to compute is that of a *cylindrical solid*, one with a constant cross-section (Figure 1). If the area of its base is A and its height is h, then its volume is $V = Ah$. This follows directly from what we mean by volume. If the base could be filled by A unit squares, then the solid could be filled by Ah unit cubes stacked on those squares.

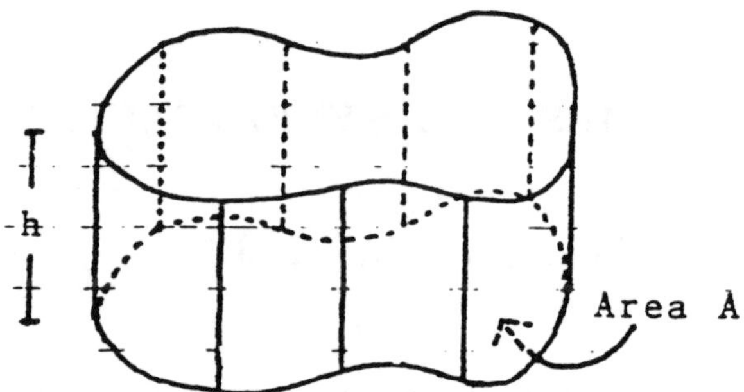

Figure 1. The volume of a cylindrical solid is $V = Ah$.

To compute the volume of a solid with variable cross-section, imagine it positioned along the x-axis, say from $x = a$ to $x = b$, as in Figure 2. The cross-section of the solid at any particular value of x has an area $A(x)$. The slicing principle says that if we know $A(x)$ for all x between a and b, then we can compute the volume by integrating $A(x)$ from $x = a$ to $x = b$.

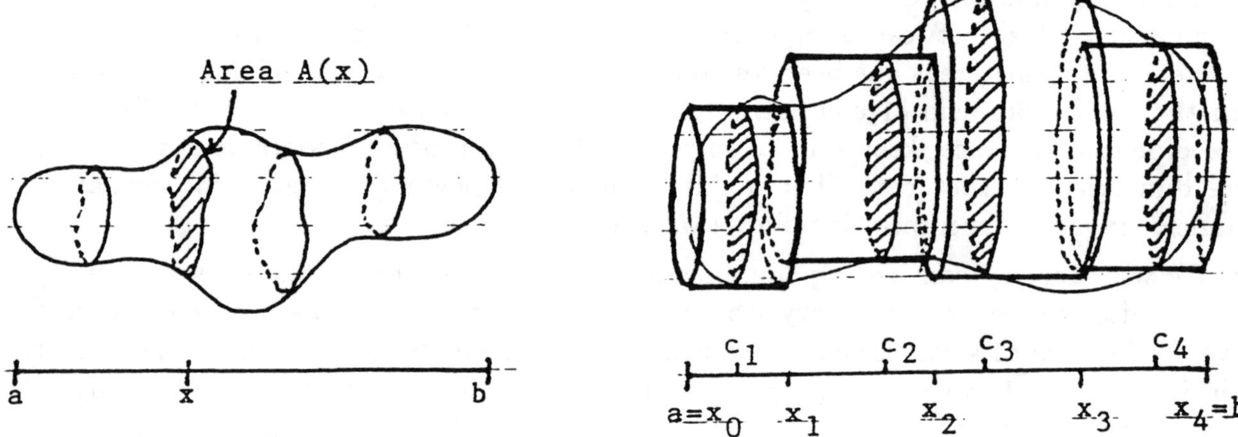

Figure 2. Slices of a volume. **Figure 3. Approximating a volume.**

The Slicing Principle. *If the cross-sectional area $A(x)$ varies continuously with x on $[a, b]$, then the volume of the solid pictured in Figure 2 is given by*

$$V = \int_a^b A(x)\, dx.$$

Proof: We partition the interval $[a, b]$ into n subintervals by points $a = x_0 < x_1 < x_2 < \ldots < x_{n-1} < x_n = b$. Let $\Delta x_i = x_i - x_{i-1}$ be the length of the ith subinterval. In each interval, choose an arbitrary point $x_{i-1} \le c_i \le x_i$. Approximate the volume above the ith subinterval by a cylindrical solid (on its side) with base area $A(c_i)$ and height Δx_i, as shown in Figure 3. The volume of this solid is the area of its base times its height, $A(c_i)\Delta x_i$. Hence the total volume of the object is approximated by the sum

$$\sum_{i=1}^{n} A(c_i)\Delta x_i,$$

which has the form of a Riemann sum. Since $A(x)$ is continuous, as $n \to \infty$ and all $\Delta x_i \to 0$ the approximation approaches the true volume, and the Riemann sum approaches the integral

$$\int_a^b A(x)\, dx,$$

and the proof is complete.

We will see how powerful a tool the slicing principle can be. One easy consequence of it is called *Cavalieri's principle*, after the 17th century Italian mathematician Bonaventura Cavalieri, who used it to compute a number of volumes in the years immediately before the development of calculus.

Cavalieri's Principle. *Suppose two solids extend along the x-axis from $x = a$ to $x = b$. Let the area of the cross-section above x be $A_1(x)$ for the first solid, $A_2(x)$ for the second solid. If $A_1(x) = A_2(x)$ for all x between a and b, then the two solids have the same volume.*

Two solids which satisfy the condition of Cavalieri's principle are called *Cavalieri congruent.*

Pyramids, Cones and Spheres

Let us use the slicing principle to verify some of those elementary school formulas. First, consider a pyramid of height h on a base which is a square of side s. To calculate its volume, put the origin at its vertex, and let the x-axis point down through its base, as in Figure 4. (Are you used to having the x-axis point "right"? Well, of course, we can put it anywhere it is useful!) The cross-sections perpendicular to the x-axis are all squares. If u is the side of the square at x, similar triangles (Figure 4) give

$$\frac{x}{u} = \frac{h}{s},$$

so that $u = xs/h$. Hence the area of the cross-section is $(xs/h)^2$, and the slicing principle says that the volume is

$$V = \int_0^h \left(\frac{xs}{h}\right)^2 dx = \frac{s^2}{h^2}\int_0^h x^2\, dx = \frac{s^2}{h^2}\left[\frac{x^3}{3}\right]_0^h = \frac{s^2}{h^2}\cdot\frac{h^3}{3} = \frac{1}{3}s^2h, \qquad (1)$$

which is, I trust, the formula you learned in elementary school. You probably didn't learn, though, a derivation of it. Did you ever wonder where the 1/3 comes from? It comes from integrating x^2.

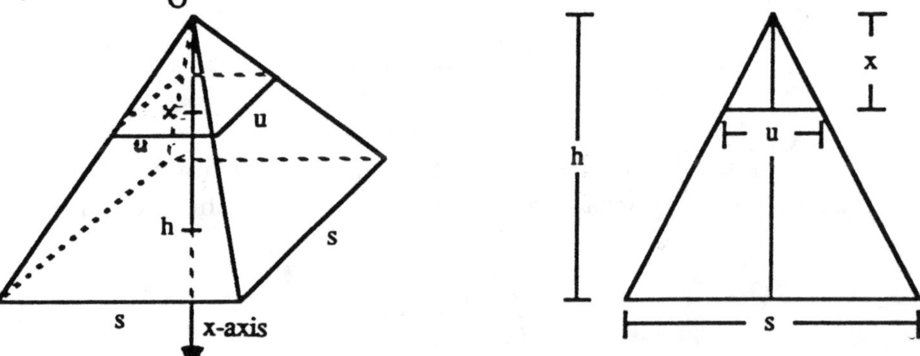

Figure 4. A pyramid, and a view through its axis.

The pyramid volume formula is one of the oldest in mathematics. It was known in Egypt by 2000 B.C. In fact, the "Moscow papyrus" of 1850 B.C. gives a formula for a more complicated shape, a truncated pyramid of height k with square "bases" of sides s and t (Figure 5). The Egyptian formula is

$$V = \frac{k}{3}\,(s^2 + st + t^2). \tag{2}$$

Did you know that one? Try deriving it in Exercise 1.

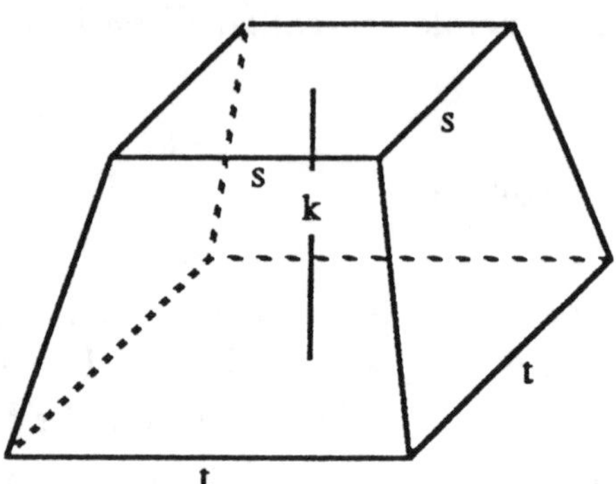

Figure 5. A truncated pyramid.

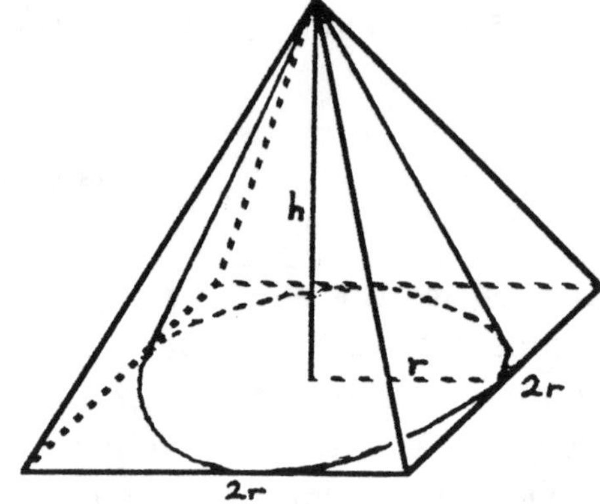

Figure 6. A cone inscribed in a pyramid.

The volume of a cone with base radius r and height h is given by

$$V = \frac{1}{3}\pi r^2 h. \tag{3}$$

This formula was discovered and proved by the Greek mathematician Eudoxus of Cnidos about 375 B.C. It follows by the same method we used for the pyramid (Exercise 2). An alternative derivation uses an argument based on Cavalieri's principle. Inscribe the cone in a pyramid whose base is a square of side $2r$ (Figure 6). The ratio of areas of any cross-section of the cone to the corresponding cross-section of the pyramid is $\pi/4$. Hence the ratio of their volumes must be $\pi/4$, so the volume of the cone is

$$V = \frac{\pi}{4}\frac{1}{3}(2r)^2h = \frac{1}{3}\pi r^2 h. \tag{4}$$

The most difficult volume problem studied by the mathematicians of Greece was determining the volume of the most "perfect" solid of all—the sphere. It was solved by the greatest of Greek mathematicians, Archimedes of Syracuse (287-212 B.C.) Let us solve the problem using calculus, and then look at what Archimedes did.

Position a sphere of radius r with its center at the origin. It lies along the x-axis from $x = -r$ to $x = r$. The cross-sections are all circles. Using the Pythagorean theorem in Figure 7a, we see that the radius of the circular cross-section at x is $y = \sqrt{r^2 - x^2}$. Hence $A(x) = \pi y^2 = \pi(r^2 - x^2)$ and the volume of the sphere is

$$V = \int_{-r}^{r} \pi(r^2 - x^2)\,dx = 2\pi \int_0^r (r^2 - x^2)\,dx = 2\pi\left[r^2x - \frac{x^3}{3}\right]_0^r = \frac{4}{3}\pi r^3. \tag{5}$$

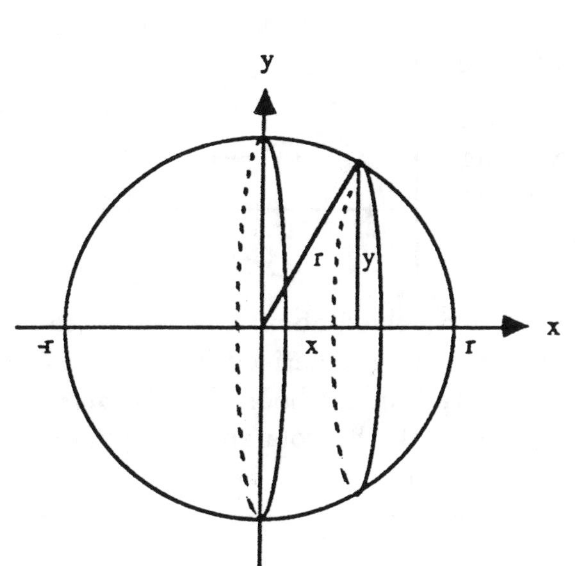

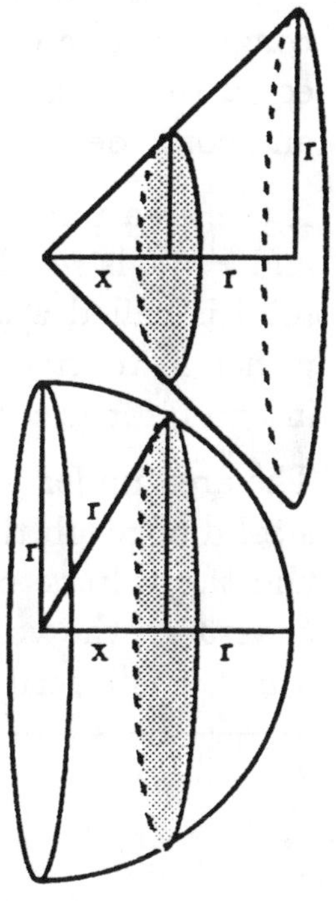

Figure 7. Finding the volume of a sphere

a. by calculus. b. as Archimedes did it.

Archimedes proved this result by the Greek "method of exhaustion" in his book *On the Sphere and the Cylinder.* It is interesting, though, that until 1906 we did not know how he discovered the result. In 1906 a copy of a previously unknown book of Archimedes, *The Method*, was discovered in Istanbul. In it, Archimedes explained that he found the volume of a sphere by something very like Cavalieri's principle. In essence, he showed that a hemisphere and a cone, positioned as in Figure 7b, are together Cavalieri congruent to a cylinder. You are asked to fill in the details in Exercise 4. In using this argument, Archimedes was 1900 years ahead of his time.

Exercises

1. Derive the Egyptian formula (2) by thinking of the truncated pyramid in Figure 5 as the difference between a large pyramid of height $h+k$ and a smaller, similar, pyramid of height h. (Draw a picture of this.) You will need to find h. Use a similar triangle argument to show that
$$\frac{h}{k} = \frac{s}{t-s}.$$
2. Use an argument similar to our argument for a pyramid to derive the formula for the volume of a cone with base radius r and height h.
3. What is the volume of a truncated cone (sometimes called the *frustrum* of a cone) with height k, bases with radii r and s? Justify your answer.
4. Show that a hemisphere and cone together, positioned as in Figure 7b, are Cavalieri congruent to a cylinder of radius r and height r. Hence the sum of their volumes must equal the volume of the cylinder. Using the formulas for the volume of the cylinder and cone, derive the formula for the volume of a sphere.
5. If the area under a graph $y = f(x)$ between $x = a$ and $x = b$ is revolved about the x-axis, the resulting solid is called a *solid of revolution.* Use the slicing principle to give a general formula, in terms of an integral, for the volume of a solid of revolution.

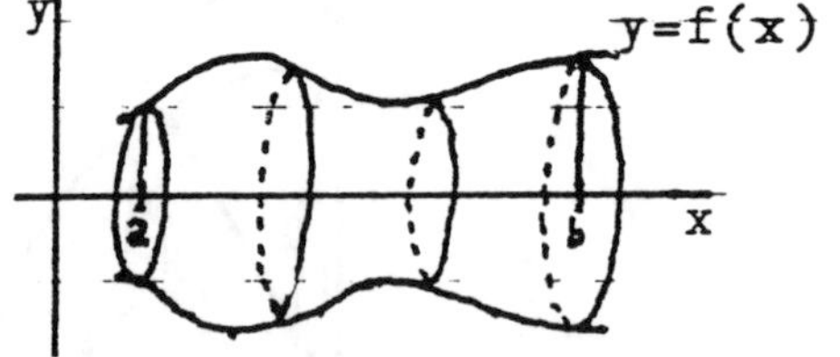

6. The area under $y = \sqrt{x}$ between $x = 0$ and $x = 4$ is revolved about the x-axis, to get a solid of revolution called a *paraboloid.* Use the formula you derived in Exercise 5 to find the volume of this paraboloid. Archimedes proved, in his book *On Conoids and Spheroids*, the volume of a paraboloid is exactly one half the volume of a cylinder in which it is inscribed. Check his result for this paraboloid.

A Modern Instance

Modern technology has produced a number of interesting shapes for which we might wish to find volumes. I bought one recently. It is a "pop-up" tent, shown in Figure 8. The base is a regular hexagon which measures 8 2/3 feet from corner to corner. Flexible exterior poles extend between opposite corners in curves that we will assume are semi-circles, to hold up the roof. How much volume does the tent enclose?

Figure 8. A pop-up tent.

To calculate the volume, we need to slice the tent (figuratively speaking) in a nice way. Slicing it vertically, either parallel to a side or parallel to a diagonal, gives interesting shapes (can you picture them?) whose areas are not easy to find. Slicing it horizontally, though, gives regular hexagons (see Figure 9). So let us start by calculating the area of a regular hexagon of "circumradius" s (see Figure 10). s is the length of hypotenuse of a 30-60-90 right triangle, whose legs must therefore have lengths $s/2$ and $\sqrt{3}s/2$. The area of this triangle is then

$$\frac{1}{2} \cdot \frac{s}{2} \cdot \frac{\sqrt{3}s}{2} = \frac{\sqrt{3}\, s^2}{8}. \tag{6}$$

The hexagon consists of 12 of these triangles, so its area must be

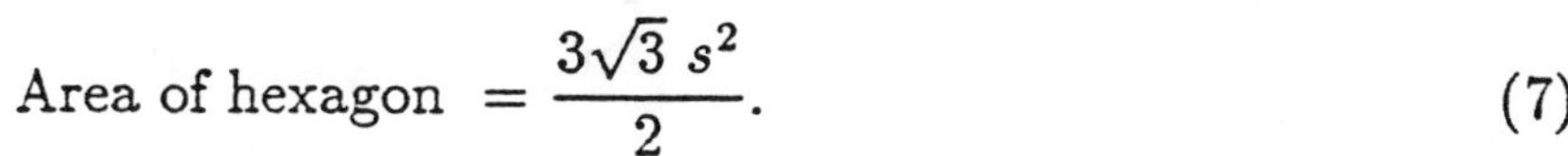

$$\text{Area of hexagon} = \frac{3\sqrt{3}\, s^2}{2}. \tag{7}$$

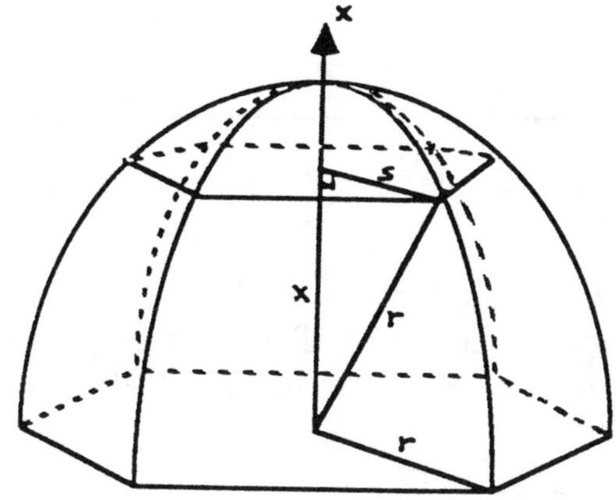

Figure 9. Slicing the tent.

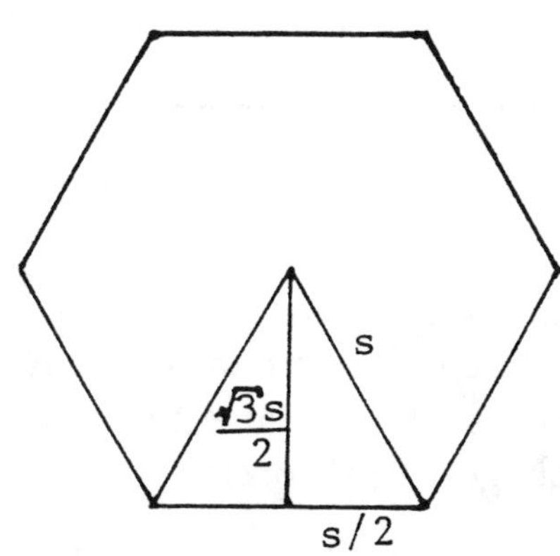

Figure 10. Area of a hexagon.

Finally, from Figure 9, if the circumradius of the base hexagon is r, the square of the circumradius of the hexagon at height x is $s^2 = r^2 - x^2$. Hence we can compute the volume of the tent,

$$\text{Volume} = \int_0^r \frac{3\sqrt{3}}{2}(r^2 - x^2)\,dx = \frac{3\sqrt{3}}{2}\left[r^2x - \frac{x^3}{3}\right]_{x=0}^{x=r} = \sqrt{3}\,r^3. \qquad (8)$$

If we had taken as our base a circle of radius r, and erected over it a hemisphere, the enclosed volume would have been $\frac{2}{3}\pi r^3$. The ratio of the volume of the tent to the volume of this enclosing hemisphere is $\sqrt{3}/\frac{2}{3}\pi \approx .827$. The tent encloses about 83% of what the hemisphere would have enclosed. On the other hand, one can make the tent out of pieces of nonstretchable fabric, which one cannot do for a hemisphere. In fact, the tent is designed to be efficient at enclosing volume, given that it is to be made out of nonstretchable fabric.

The answer to our original problem, by the way, is that the volume of my new tent is $\sqrt{3}(13/3)^3 \approx 141$ cubic feet. That's pretty good for something which weighs only seven pounds. The exercises give you a chance to try computing some other interesting volumes.

Exercises

7. a) Compute the volume of the ring which remains after a hole of radius b is bored through a sphere of radius r.

   b) Show the the volume depends only on the height h of the ring, not on the specific values of b and r.

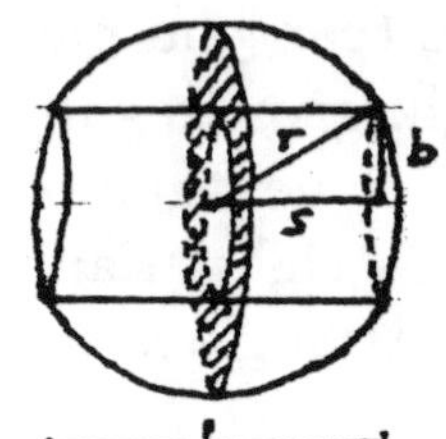

8. Compute the volume of the perpendicular intersection of two cylindrical pipes of radius r. [Hint: if you slice it right, the cross-sections are squares.] This volume was computed by Archimedes, and also by the Chinese mathematician Tsu Ch'ung-Chih.

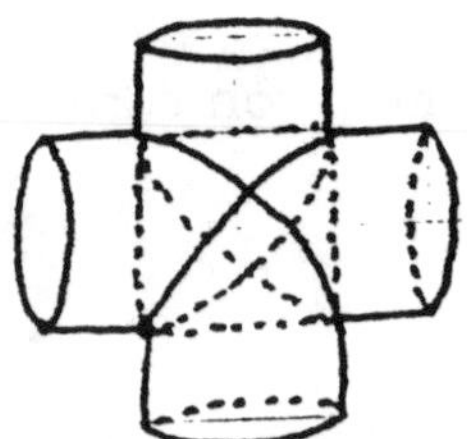

Volumes in Higher Dimensions

To a mathematician, higher dimensional spaces are not mysterious, at least in their definition. Three dimensional space R^3 can be thought of as the set of all triples (x, y, z) of real numbers. To measure the distance between two points (x, y, z) and (x', y', z'), we use the Pythagorean theorem:

$$\text{dist}((x,y,z),(x',y',z')) = \sqrt{(x-x')^2 + (y-y')^2 + (z-z')^2}.$$

In an exactly analogous way, 4-dimensional space R^4 can be thought of as the set of all 4-tuples (x, y, z, w) of real numbers, with

$$\text{dist}((x,y,z,w),(x',y',z',w')) = \sqrt{(x-x')^2 + (y-y')^2 + (z-z')^2 + (w-w')^2}. \qquad (9)$$

Five and higher dimensional spaces are handled in exactly the same way. For a while, though, let's stay in four dimensions. In four dimensions, just like in three, a sphere of radius r about the origin $O = (0,0,0,0)$ is the set of all points at distance r from O. Hence, given the definition of distance in R^4, the equation of a sphere is

$$x^2 + y^2 + z^2 + w^2 = r^2. \tag{10}$$

Science fiction writers often call spheres in higher dimensions *hyperspheres*, and talk about their *hypervolume*. In fact, I used that terminology in the title of this module, thinking that you might have heard it. However, mathematicians usually omit the "hyper", and just talk about spheres and volume in higher dimensional spaces. Mathematicians also use terminology which may be slightly different from colloquial usage when they distinguish between the *sphere* $x^2+y^2+z^2+w^2 = r^2$ and the 4-dimensional *ball* $x^2+y^2+z^2+w^2 \leq r^2$. Thus, what we are about to do is find the volume of a 4-dimensional ball.

I will use the notation $V_n(r)$ for the n-dimensional volume of a ball of radius r in n-dimensional space. For example, $V_2(r) = \pi r^2$, the 2-dimensional volume (i.e. area) of a ball (i.e. disk) of radius r in 2-dimensional space. We have seen that $V_3(r) = \frac{4}{3}\pi r^3$.

What is the volume, $V_4(r)$, of a 4-dimensional ball of r? We will find the answer by using the slicing principle, one dimension up. (Should we call it the "hyperslicing principle"?). The 4-D ball of radius r is laid out along the x-axis between $x = -r$ and $x = r$. For a fixed value of x, the slice above x is the set of all (x, y, z, w) with that value of x, and $y^2 + z^2 + w^2 \leq r^2 - x^2$. This is a 3-dimensional ball of radius $\sqrt{r^2 - x^2}$, and its volume is

$$V_3(\sqrt{r^2 - x^2}) = \frac{4}{3}\pi\left(\sqrt{r^2 - x^2}\right)^3 = \frac{4}{3}\pi(r^2 - x^2)^{3/2}. \tag{11}$$

Thus by the slicing principle, the volume of the 4-dimensional ball of radius r is

$$\begin{aligned} V_4(r) &= \int_{-r}^{r} \frac{4}{3}\pi(r^2 - x^2)^{3/2} dx \\ &= \frac{8}{3}\pi \int_0^r (r^2 - x^2)^{3/2}\, dx. \end{aligned} \tag{12}$$

There is a standard method for attacking integrals involving $\sqrt{r^2 - x^2}$, which is to introduce a new variable by writing $x = r\sin\theta$. Then

$$\begin{aligned} \sqrt{r^2 - x^2} &= r\sqrt{1 - \sin^2\theta} = r\cos\theta \\ dx &= r\cos\theta\, d\theta \\ \text{when } x &= 0, \quad \theta = 0 \\ \text{when } x &= r, \quad \theta = \pi/2. \end{aligned}$$

With this substitution, we get

$$V_4(r) = \frac{8}{3}\pi r^4 \int_0^{\pi/2} \cos^4\theta\, d\theta. \tag{13}$$

The nicest way I know to do that integral is by using an integral reduction formula which says that

$$\int_0^{\pi/2} \cos^n \theta \, d\theta = \frac{n-1}{n} \int_0^{\pi/2} \cos^{n-2} \theta \, d\theta \qquad (n \geq 2). \tag{14}$$

If you know the technique called "integration by parts," you are invited to prove this elegant formula in Exercise 10. Using it, we get

$$\int_0^{\pi/2} \cos^4 \theta \, d\theta = \frac{3}{4} \int_0^{\pi/2} \cos^2 \theta \, d\theta = \frac{3}{4} \cdot \frac{1}{2} \int_0^{\pi/2} \cos^0 \theta \, d\theta = \frac{3}{4} \cdot \frac{1}{2} \cdot \frac{\pi}{2} = \frac{3\pi}{16}, \tag{15}$$

and hence our answer for the volume of a 4-dimensional ball of radius r is

$$V_4(r) = \frac{8\pi r^4}{3} \cdot \frac{3\pi}{16} = \frac{\pi^2}{2} r^4. \tag{16}$$

Would you have guessed it? I catch myself wishing that Archimedes might be brought back just for a moment, for the pleasure of seeing this calculation.

Of course, there is no reason to stop at dimension four. The slicing formula

$$V_n(r) = \int_{-r}^{r} V_{n-1}\left(\sqrt{r^2 - x^2}\right) dx \tag{17}$$

works in general, so that having calculated the volume of an $(n-1)$-dimensional ball, we can proceed to calculate the volume of an n-dimensional ball. Let's follow the argument a little further, because something interesting emerges.

The volume of an n-dimensional ball of radius r has the form

$$V_n(r) = u_n r^n$$

where u_n is the volume of a ball of unit radius. You already know the first few u_n's:

$u_1 = 2$ (A 1-dimensional ball is a line segment, right?)

$$u_2 = \pi$$

$$u_3 = \frac{4\pi}{3}$$

$$u_4 = \frac{\pi^2}{2}$$

The slicing formula and the kind of work we did above enable us to use u_{n-1} to find u_n:

$$u_n = 2u_{n-1} \int_0^{\pi/2} \cos^n \theta \, d\theta, \qquad n = 1, 2, \ldots \tag{18}$$

Furthermore, using the integral reduction formula repeatedly, as we did in (15), yields

$$\int_0^{\pi/2} \cos^n \theta \, d\theta = \begin{cases} \frac{(n-1)(n-3)\cdots 3\cdot 1}{n(n-2)\cdots 4\cdot 2} \cdot \frac{\pi}{2} & \text{if } n \text{ is even} \\ \frac{(n-1)(n-3)\cdots 4\cdot 2}{n(n-2)\cdots 5\cdot 3} \cdot 1 & \text{if } n \text{ is odd.} \end{cases} \tag{19}$$

Using (18) and (19), you can check that

$$u_5 = \frac{8\pi^2}{15} \qquad u_6 = \frac{\pi^3}{6} \qquad u_7 = \frac{16\pi^3}{105}$$

and continue as far as you want. If you are ambitious, you can even derive a general formula for u_n (Exercise 11).

The interesting facts I want to point out emerge when we tabulate the u_n's numerically:

n	u_n	$u_n/2^n$
1	2.000	1.000
2	3.142	0.785
3	4.189	0.524
4	4.935	0.308
5	5.264	0.164
6	5.168	0.081
7	4.725	0.037
8	4.059	0.016
9	3.299	0.006
10	2.550	0.002

...

First, notice that the n-dimensional volume of a unit ball increases up through dimension 5, but then begins to decrease, and in fact approaches 0 as n approaches ∞. It makes you wonder: why should a unit ball be biggest in dimension 5? Second, consider $u_n/2^n$, which is the ratio of the volume of a unit ball and the smallest cubical box (side 2) which would enclose it. That ratio goes to zero monotonically and rapidly as n increases. As dimension increases, a ball gets curved in more directions, and takes up an ever smaller percentage of the volume of a box in which it is tightly packed. A 10-dimensional ball only takes up about 1/400 of the smallest 10-dimensional cube which contains it.

Exercises

9. Carry out in detail the calculation for the volume of a 5-dimensional ball of radius r, as we did above for a 4-dimensional ball.

10. If you know the integration by parts formula

$$\int_a^b u\,dv = uv\Big|_a^b - \int_a^b v\,du,$$

use it with $u = \cos^{n-1}\theta$, $dv = \cos\theta\,d\theta$ to verify the reduction formula (14). [You will also need to use the identity $\sin^2\theta = 1 - \cos^2\theta$.]

11. a) Use (18) and (19) to verify the values given for u_5, u_6 and u_7.

*b) Use (18) and (19) twice to get an expression for u_n in terms of u_{n-2}. Use that expression to derive the following formulas for the u_n's:

$$u_{2k} = \frac{\pi^k}{k!}$$
$$u_{2k+1} = \frac{2^{k+1}\pi^k}{1 \cdot 3 \cdot 5 \cdots (2k+1)}.$$

Uses of Higher Dimensions

I hope you enjoyed our excursion into volume in higher dimensions, perhaps even enough that you never asked, "What good is it?" But what good is it, when we live in a space of only three dimensions? It turns out that higher dimensional models appear regularly in all areas of modern science. Here are three simple illustrative examples.

1) The state of a particle moving in three dimensional space can be described by six numbers $(x_1, x_2, x_3, v_1, v_2, v_3)$, the first three giving its location in space and the last three giving its velocity in each of the three coordinate directions. In other words, the state is a point in R^6. If we confine the particle to a bounded region in R^3 and also restrict its velocity, the state is confined to a bounded region in R^6. Sometimes it is important to know the volume of that region.

2) Suppose I give you four line segments whose lengths x, y, z, w are randomly chosen from $[0,1]$. What is the probability that you will be able to make a quadrilateral from the segments? This probability question can be answered by computing the volume of the set of points (x, y, z, w) in the unit cube in R^4 which satisfy the constraints

$$x+y+z \geq w \qquad x+y+w \geq z \qquad x+z+w \geq y \qquad y+z+w \geq x.$$

Probability questions often reduce to questions of volumes in high dimensions. (By the way, the above probability turns out to be 5/6.)

3) In satellite communication, information is sent in *binary words*, strings of energy pulses indicating zeros and ones, such as 10010. Because of interfering "noise", what is received may look like $a_1a_2a_3a_4a_5$, where each a_i is a real number between 0 and 1. Suppose we agree to decode such a string as the binary word $e_1e_2e_3e_4e_5$ if and only if

$$\text{dist}((a_1, a_2, a_3, a_4, a_5), (e_1, e_2, e_3, e_4, e_5)) < 1/2$$

in R^5. Decoding will then be non-ambiguous. On the other hand, since the points within distance 1/2 of a corner of the unit 5-cube fit together to make a 5-dimensional ball of radius 1/2, we will only be able to decode 16.4% of possible received words. If our codewords had length 10, we would be able to decode 0.2% of possible received words.

Banchoff [1990] contains many examples of real uses of higher dimensional volumes. It also contains beautiful computer generated pictures of higher dimensional objects—highly recommended! If you like working with higher dimensional spheres, Fraser [1984] does more of it, in context of a light-hearted problem.

References

Aaboe, Asger (1964), *Episodes from the Early History of Mathematics*, New Mathematical Library #13, Mathematical Association of America. Chapter 3 discusses Archimedes' life and works, and his calculation of the volume of a sphere.

Archimedes (about 250 B.C.), *The Works of Archimedes, with the 'Method' of Archimedes*, edited by T.L. Heath, Dover.

Banchoff, Thomas (1990), *Beyond the Third Dimension*, Scientific American Library.

Eves, Howard (1991), "Two surprising theorems on Cavalieri congruence," *College Mathematics Journal* 22: 118-124.

Fraser, Marshall (1984), "The grazing goat in n dimensions," *College Mathematics Journal* 15: 126-134.

Answers to Exercises

1.

$$\begin{aligned} V &= \frac{1}{3}t^2(h+k) - \frac{1}{3}s^2h \\ &= \frac{1}{3}\left[t^2\left(\frac{ks}{t-s}+k\right) - s^2\frac{ks}{t-s}\right] \\ &= \frac{k}{3(t-s)}[t^2(s+(t-s)) - s^3] \\ &= \frac{k}{3}\cdot\frac{t^3-s^3}{t-s} = \frac{k}{3}[t^2+st+s^2]. \end{aligned}$$

2.

$$V = \int_0^h \pi\left(\frac{xr}{h}\right)^2\,dx = \frac{1}{3}\pi r^2 h.$$

3. $V = \frac{\pi k}{3}(r^2+rs+s^2)$. The easy way to justify this is to compare the truncated cone in a truncated pyramid, whose volume you know by the Egyptian formula, appealing to Cavalieri's principle as we did for the cone.

4.

$$\begin{aligned} A(x) &= (\text{slice of cone}) + (\text{slice of hemisphere}) = \pi x^2 + \pi(r^2-x^2) = \pi r^2 \\ &= (\text{slice of cylinder}). \end{aligned}$$

5.

$$V = \int_a^b \pi(f(x))^2\,dx.$$

6.

$$V = \int_0^4 \pi x\,dx = \pi\frac{x^2}{2}\bigg|_0^4 = 8\pi.$$

The volume of the circumscribing cylinder is $\pi 2^2\cdot 4 = 16\pi$.

7.

$$\begin{aligned} V &= 2\int_0^s \left(\pi(r^2-x^2)-\pi b^2\right)\,dx = 2\pi\left[(r^2-b^2)x - \frac{x^3}{3}\right]_0^s \\ &= 2\pi\left(s^3-\frac{s^3}{3}\right) = \frac{4\pi s^3}{3} \qquad \text{since } r^2-b^2=s^2 \\ &= \frac{\pi h^3}{6} \qquad \text{since } h = \frac{s}{2}. \end{aligned}$$

8. If we slice horizontally, the cross-sections are squares. The square at height has side $2\sqrt{r^2-x^2}$. Hence

$$V = \int_{-r}^r (2\sqrt{r^2-x^2})^2\,dx = 8\int_0^r (r^2-x^2)\,dx = \frac{16r^3}{3}.$$

9.

$$V_5(r) = 2\int_0^r \frac{\pi^2}{2}\left(\sqrt{r^2 - x^2}\right)^4 dx = \pi^2 \int_0^r (r^4 - 2r^2x^2 + x^4)\, dx$$
$$= \pi^2 \left[r^4 x - \frac{2}{3}r^2x^3 + \frac{x^5}{5}\right]_0^r = \frac{8\pi^2}{15}.$$

10.

$$\int_0^{\pi/2} \cos^{n-1}\theta \cos\theta \, d\theta = \cos^{n-1}\theta \sin\theta\Big|_0^{\pi/2} - \int_0^{\pi/2} \sin\theta \cdot (n-1)\cos^{n-2}\theta(-\sin\theta)\, d\theta$$
$$= (n-1)\int_0^{\pi/2} \cos^{n-2}\theta \, d\theta - (n-1)\int_0^{\pi/2} \cos^n \theta \, d\theta$$

since the non-integral term evaluates to zero when $n \geq 2$, and using $\sin^2\theta = 1 - \cos^2\theta$. Now bring the last term to the left side of the equation, and divide both sides by n.

11. a) For example

$$u_5 = 2u_4 \int_0^{\pi/2} \cos^5\theta \, d\theta = 2\frac{\pi^2}{2}\frac{4\cdot 2}{5\cdot 3} = \frac{8\pi^2}{15}.$$

b)

$$u_n = 2\frac{(n-1)(n-3)\cdots}{n(n-2)\cdots} \cdot 2\frac{(n-2)(n-4)\cdots}{(n-1)(n-3)\cdots} \cdot \frac{\pi}{2} u_{n-2} = \frac{2\pi}{n} u_{n-2}.$$

The formula for odd n follows directly. The formula for even n requires observing that

$$\frac{2^k}{2\cdot 4 \cdots (2k)} = \frac{1}{k!}.$$

RELIABILITY AND THE COST OF GUARANTEES

Author: Kevin J. Hastings, Knox College, Galesburg, IL 61401

Areas of Application: business, manufacturing

Calculus needed: definition of derivative, techniques of differentiation, Fundamental Theorem of Calculus, integration by parts and substitution, Riemann sums, exponential functions.

Related mathematics: continuous probability distributions

Satisfaction Guaranteed...Or Your Money Back!

Frequently we buy items that come with guarantees. Your video cassette recorder carries a year of free repairs, your new car has 30,000 miles on the drive train, that incredible food processor you ordered from the late night television ad will give total satisfaction "or your money back." The companies that offer these guarantees are on occasion forced to pay for their promises. Since they are in business to make money, they will have to raise the price of the item in order to balance the risk they are taking by issuing such contracts on imperfectly reliable equipment. In this module we are interested in determining how they can decide on an appropriate pricing scheme in the presence of the possible costs when the piece of equipment fails.

To state a particular problem, the author recently purchased a set of steel-belted radial tires from Sears, Roebuck for $75 each. The tires came with a limited warranty that included the following rule about tread wearout.

> "Until 50,000 miles of usage...Sears will, at its option, either replace the tire charging the proportion of the current price that represents the ratio of miles usage actually received to the number of miles covered by the warranty or give a refund to the proportion off the purchase price that represents the ratio of miles remaining to the number of miles covered by the warranty if wearout of the tread (2/32″) occurs. (Sears Passenger and Light Truck Tire Warranty 14875 Rev. 9/87)"

This is stated in a way that is not easy to read, but what it seems to say is this: if my tread fails at a mileage T (in thousands of miles) prior to 50 thousand miles, the company must give to me material or cash equivalent that is worth a certain proportion of the original price p. This proportion depends on the unused part of my 50 thousand miles. Specifically, they refund to me a total of

$$p\,\frac{50-T}{50} = p\left(1-\frac{T}{50}\right) \tag{1}$$

dollars, in case the tread failure mileage T is less than 50, and nothing otherwise.

Let p_1 be the total revenue the company would like to receive from selling this tire. That is, p_1 is the cost of making the tire, plus the desired profit margin. Let w be the surcharge that will be necessary to compensate for the warranty. Then the total price of the tire will be set at $p = p_1 + w$. Taking the role of the company, we assume that we know p_1, and we want to determine w. Then

$$\begin{aligned} p_1 &= \text{ selling price } - \text{ average refund due to warranty} \\ &= (p_1 + w) - (p_1 + w) \cdot (\text{average of } 1 - T/50 \text{ when } T < 50). \end{aligned} \tag{2}$$

We would like to solve for w, but here we run into a difficulty which we will explore in detail in this module. What do we mean by the "average of $1 - T/50$ when $T < 50$", and how do we compute it? There is inherent randomness in the quantity T, the mileage at tread failure. There is no way of predicting perfectly how long a given tire will survive.

A representative sample of 25 tire lifetimes, sorted into increasing order, is below:

Tire Lifetimes (in thousands of miles)

21.39	33.19	37.91	44.98	58.32
24.16	33.34	38.63	45.19	58.52
26.37	33.37	39.06	45.83	59.00
29.49	35.82	41.20	49.94	59.98
31.94	36.70	43.32	56.43	65.29

Most of the tires do fail before the 50 thousand mile mark, but this causes no great problem for the company if they set a high enough surcharge. The goal is to take the data set, form a model of the random behavior of failures, and compute the average value mentioned above so as to determine the proper surcharge.

Probability Distributions

To make progress toward formulating mathematical models for reliability problems, we must discuss some definitions and concepts from elementary probability. In this section you will learn about (a) randomness; (b) probability; (c) random variables; (d) probability distributions; and (e) conditional probability.

A probabilist who uses the word 'random' in reference to some experiment or phenomenon means that there may be one, several, countably many, or even a continuum of possible outcomes, and that the exact outcome will not be known until the experiment is performed or the phenomenon is observed. In English, the word 'random' is occasionally used casually with the unfortunate connotation of 'utterly unknown.' This is not the case in probability, where we usually assume that we know something about the likelihood of occurrence of outcomes. For example, if we roll two dice, there are 36 possible outcomes, only one of which is the combination (1,1). Hence, if we are willing to assume that the dice are fairly weighted, the likelihood that both faces will come up as 1 is 1/36. Similarly, the likelihood that the sum of the up faces is 3 should be 2/36, since the two outcomes (1,2) and (2,1) which produce this event each have likelihood 1/36. The likelihood that the sum of the faces will be less than or equal to 12 is 1 or 100%, since all 36 possible outcomes result in the occurrence of this event. The likelihood that the sum is 14 is 0, since no outcomes permit this to happen.

The preceding illustration introduces the idea of *probability* of outcomes, and of sets of outcomes, which are called *events*. We will denote the probability that an event A occurs by $P[A]$. The reader may consult any of the references on probability listed at the end for more detail, but for our purposes the following are the most important, and intuitively obvious properties of probability:

$$0 \le P[A] \le 1, \tag{3}$$

$$P[A \text{ or } B] = P[A] + P[B] \text{ if } A \text{ and } B \text{ have no outcomes in common,} \tag{4}$$

$$P[A] + P[A^c] = 1, \text{ where } A^c \text{ is the event "not } A\text{."} \tag{5}$$

The first property says that probabilities, as measures of likelihood, should be between 0% and 100%. The probability of the entire set of possible outcomes is one, and the probability of an empty set of outcomes is zero. The second property says that the total probability of an event is the sum of the probabilities of the disjoint events of which it is composed. The third property says that it is certain that either A will occur or it will not occur .

Frequently in applications of probability theory, the events of interest involve random numerical-valued quantities called *random variables*. For instance, consider a device such as a battery which functions during a certain interval of time, and then fails. The time T of failure is a random variable, whose possible values are real numbers in the interval $[0, \infty)$. We might be interested in probabilities such as $P[T \le t]$, i.e. the probability that the failure time of the battery occurs at time t or before.

If T is a random variable, then the function

$$F(t) = P[T \le t] \tag{6}$$

is called the *cumulative distribution function* (c.d.f.) of T. From the cumulative distribution function, probabilities such as the following can be calculated:

$$P[T > t] = 1 - F(t), \tag{7}$$

$$P[a < T \le b] = P[T \le b] - P[T \le a] = F(b) - F(a). \tag{8}$$

The first line uses property (5) of probability, and the second line follows from property (4).

In the context of the tire example, one way of statistically estimating the c.d.f. of tire lifetime T is to test some number N of tires, order their failure times (say $T_1,\ T_2,\ \ldots\ , T_N$), and plot a step function which for each time gives the proportion of tires that have failed by that time. This function, called an *empirical c.d.f.*, takes the value 0 at $t = 0$, and jumps up by $1/N$ at each failure time, that is:

$$F_N(t) = \begin{cases} 0 & \text{if } t < T_1; \\ k/N & \text{if } T_k \le t < T_{k+1}, k = 1, ..., N-1; \\ 1 & \text{if } t \ge T_N. \end{cases} \tag{9}$$

A typical graph of an empirical distribution function, together with the c.d.f. for which it is an estimate, is shown in Figure 1. It is based on a sample of five random observations (following the exponential distribution with parameter 1, described below), whose ordered values are .419, .570, 1.08, 1.15, and 1.64.

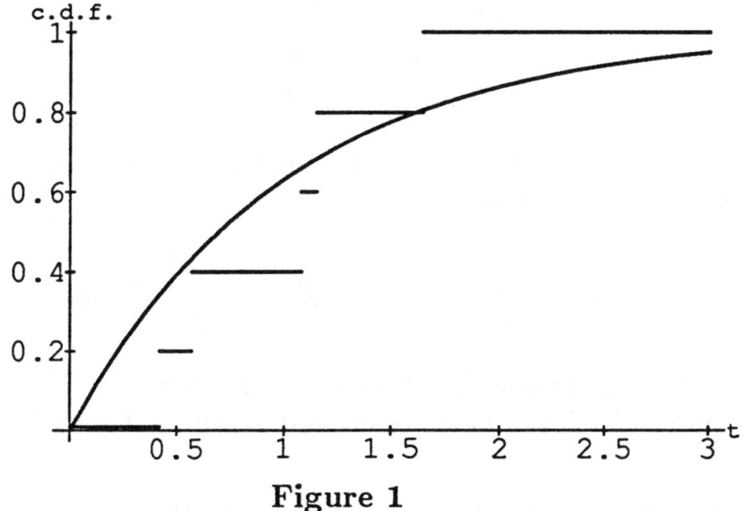

Figure 1

Another perspective is often taken on the calculation of probabilities such as those in (6), (7), and (8). Suppose that the c.d.f. F is differentiable with derivative $F'(t) = f(t)$. Then for small Δt,

$$\frac{F(t + \Delta t) - F(t)}{\Delta t} \approx f(t). \tag{10}$$

Moving Δt to the right, and using formula (8),

$$P[t < T \le t + \Delta t] = F(t + \Delta t) - F(t) \approx f(t)\Delta t. \tag{11}$$

The function $f(t)$ is called the *probability density function* (p.d.f.) of the random variable T, and we see from the above equation that it measures the rate at which probability is accumulating at the point t. A typical cumulative distribution function F and its derivative, the density function f, are pictured below. Because $f = F'$, the distribution function will be concave up when the density function is increasing and concave down when the density function is decreasing.

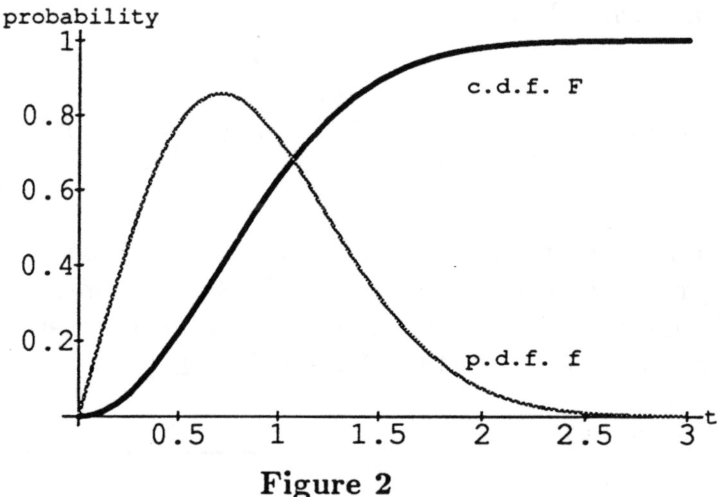

Figure 2

By the Fundamental Theorem of Calculus, probabilities can be calculated by computing areas under the density function:

$$P[a < T \leq b] = F(b) - F(a) = \int_a^b f(t)\, dt. \tag{12}$$

For instance, if T has the density shown as the light curve in Figure 2, the probability that T lies in the interval $[.5, 1.5]$ is the area under the light curve and above the x-axis between .5 and 1.5.

Two particular distributions studied in reliability theory and other applications of probability are:

1. Exponential distribution with parameter $\lambda > 0$:

$$\text{p.d.f. } f(t) = \begin{cases} \lambda e^{-\lambda t} & \text{if } t \geq 0; \\ 0 & \text{otherwise.} \end{cases} \tag{13}$$

2. Weibull distribution, parameters $\lambda > 0$ and $\beta > 0$:

$$\text{c.d.f. } F(t) = \begin{cases} 1 - e^{-(\lambda t)^\beta} & \text{if } t \geq 0; \\ 0 & \text{otherwise.} \end{cases} \tag{14}$$

The exponential distribution has been known to be a good model for the distribution of lifetimes of electrical components (Davis, 1952) and other objects that do not age appreciably, but are subject to rare catastrophic events. The Weibull distribution is probably the most frequently used of all distributions in modelling lifetimes of mechanical devices, partly because the empirical distribution functions of samples of device lifetimes tend to fit the c.d.f. (14) (see, e.g. (Barlow and Proschan, 1975, p. 73) on ball bearings and vacuum tubes). There are theoretical reasons to choose this distribution as well. A standard result in reliability theory (Gertsbakh, 1989, p. 26) implies that a system of many highly reliable components with exponential lifetimes possesses a system lifetime which has approximately the Weibull distribution.

Exercises

1. Show that the c.d.f. of the exponential distribution is

$$F(t) = \begin{cases} 1 - e^{-\lambda t} & \text{if } t \geq 0; \\ 0 & \text{otherwise.} \end{cases} \tag{15}$$

2. Show that the p.d.f. of the Weibull distribution with parameters λ and β is

$$f(t) = \begin{cases} \lambda^{\beta} \beta t^{\beta-1} e^{-(\lambda t)^{\beta}} & \text{if } t \geq 0; \\ 0 & \text{otherwise.} \end{cases} \tag{16}$$

Why does it not matter from a probabilistic point of view how f is defined at $t = 0$?

3. Suppose that a microprocessor has a lifetime which is exponentially distributed with parameter $\lambda = .001$. Find in two ways the probability that the processor survives at least until time 1000.

4. Another p.d.f. that arises in reliability theory as the distribution of the lifetime of a system with an exponential component that has several backups is the *gamma distribution.* One instance of it is the following density function:

$$f(t) = \begin{cases} 4te^{-2t} & \text{if } t \geq 0; \\ 0 & \text{otherwise.} \end{cases} \tag{17}$$

Show that the c.d.f. of this density is:

$$\text{c.d.f. } F(t) = \begin{cases} 1 - e^{-2t} - 2te^{-2t} & \text{if } t \geq 0; \\ 0 & \text{otherwise.} \end{cases} \tag{18}$$

Compute the probability that a device with this lifetime distribution fails in the time interval $[1, 5]$.

*5. If you have a graphics package available, do the following. Plot the exponential p.d.f. and its c.d.f. for $\lambda = \frac{1}{3}, \frac{1}{2}, 1, 2,$ and 3. How does the shape change as λ increases? What is the significance of this to the lifetime of devices? Repeat the exercise for the Weibull p.d.f. and its c.d.f. for the same values of λ, and each of the values $\beta = \frac{1}{2}$ and $\beta = 2$.

The last background item that we need to introduce is the idea of ***conditional probability.*** If one event is known to have occurred, this fact might influence the probability of occurrence of another, since it is no longer the complete set of outcomes that matters, but only those in the known event. For example, if it is known that a device has already lasted until time .5, then the probability that the device works until time 1.5 should be calculated as if the space of possible outcomes were the set of times $t > .5$. Viewing these probabilities as areas under the density curve $f(t)$, the conditional probability that $T > 1.5$ given that $T > .5$ should be defined as the ratio of the area to the right of 1.5 to the entire area lying to the right of the point $t = .5$ (see Figure 3). In other words,

$$P[T > 1.5 \mid T > .5] = \frac{\int_{1.5}^{\infty} f(t)dt}{\int_{.5}^{\infty} f(t)dt} = \frac{1 - F(1.5)}{1 - F(.5)}, \tag{19}$$

where the symbol "|" is read "given that."

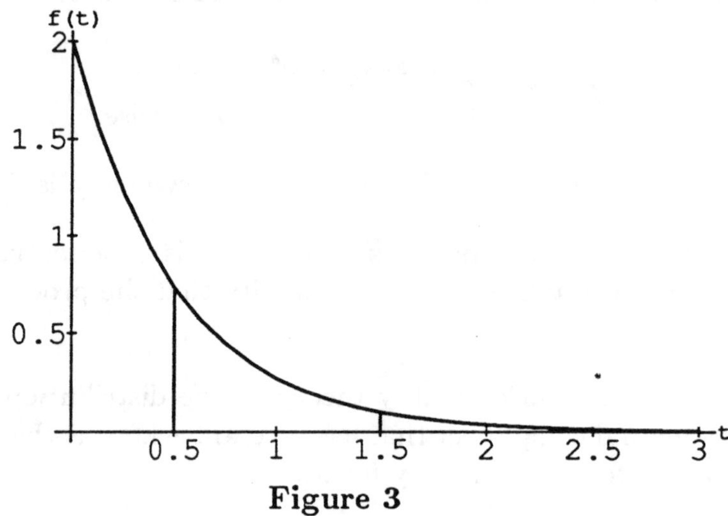

Figure 3

In this special case, the event $T > 1.5$ happens to be a subset of the event $T > .5$, so that all its outcomes also belong to the event that is known. In the general definition of the conditional probability of an event B listed below, we explicitly require that the outcomes which B may have that lie outside the known event do not contribute to its conditional probability.

Definition. *The conditional probability of an event B, given an event A, denoted by $P[B|A]$, is defined by*

$$P[B|A] = \frac{P[A \text{ and } B]}{P[A]}, \tag{20}$$

when the denominator is not zero.

Thus, B is given probability according to its relative likelihood within the event A.

Exercise

6. For the microprocessor in Exercise 3, find $P[T > 1000 \mid T > 500]$.

Reliability and Failure Rate

We consider systems which work initially, and fail at some random time T. The *reliability* of the system is the function of t which gives the probability that the system still works at time t, i.e. its lifetime exceeds t:

$$R(t) = P[T > t] = 1 - F(t). \tag{21}$$

Formula (14) implies that

$$R(t) = e^{-(\lambda t)^{\beta}} \tag{22}$$

is the reliability function for the Weibull distribution. Several instances of this function, with $\lambda = 2$, and $\beta = .5, 1, 1.5$ are illustrated in Figure 4, in which the dashed line is the case $\beta = .5$, the solid line is the case $\beta = 1$, and the bold line is the case $\beta = 1.5$. The rise in the β value does produce a fall in reliability after some time elapses, but the real qualitative differences between these distributions are not yet clear. We will better understand these differences shortly, when we define the idea of failure rate.

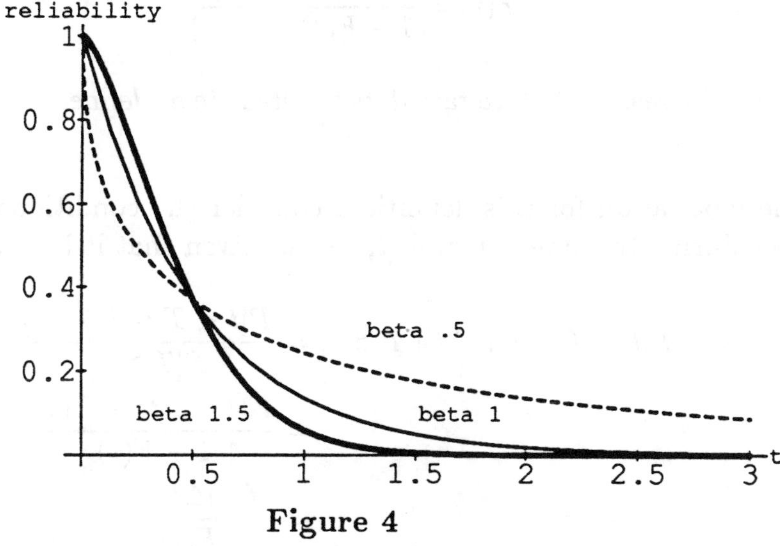

Figure 4

Just as the c.d.f. of a distribution can be estimated by a sample empirical distribution function, the reliability function can be estimated by the *empirical reliability function.* The value of this function at time t is the proportion of items in the sample that lived beyond t; hence it must equal one minus the value of the empirical c.d.f. at time t. Figure 5 shows the empirical reliability function for the tire data set. The horizontal axis is in units of thousands of miles.

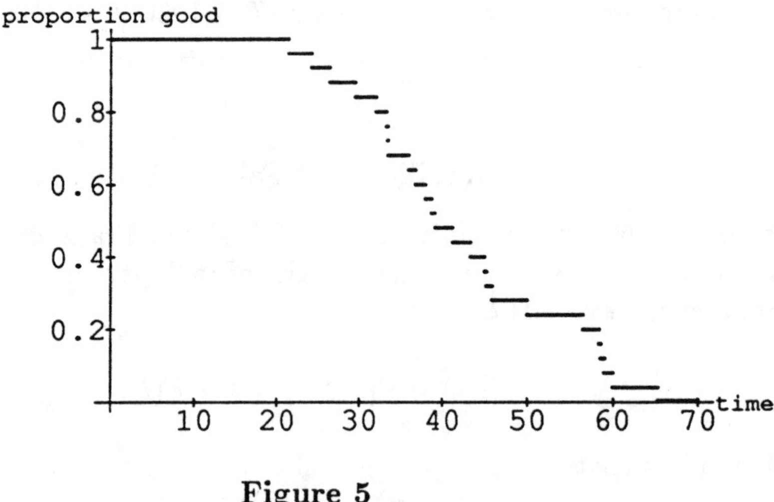

Figure 5

Intuitively, one would think that most physical devices are subject to wear, and that as they age, they become more and more likely to fail. To form a mathematical model for this aging process, we introduce the following definition.

Definition. *The failure rate of a distribution F with density f is defined by:*

$$h(t) = \frac{f(t)}{1-F(t)} = \frac{f(t)}{R(t)}. \tag{23}$$

A distribution has increasing failure rate if its failure rate h, defined by (23), is an increasing function of t.

To see the motivation for this definition, consider the conditional probability that a component fails during the time interval $(t, t+\Delta t]$ given that it has survived until time t:

$$\begin{aligned} P[t < T \le t+\Delta t \mid T > t] &= \frac{P[t < T \le t+\Delta t]}{P[T > t]} \\ &= \frac{F(t+\Delta t) - F(t)}{1-F(t)} \\ &\approx \frac{f(t)\Delta t}{1-F(t)} \\ &= h(t)\Delta t. \end{aligned} \tag{24}$$

Dividing both sides by Δt and letting $\Delta t \to 0$ shows that $h(t)$ is indeed the rate at which failure probability accumulates at time t, given that the device has survived until time t. The IFR property may be interpreted as an aging condition as follows: the longer a device survives, the greater is its probability of failure in the next infinitesimally small interval of time.

For example, consider the Weibull distribution with $\lambda = 1$ and $\beta = 2$. From (16) and (22) we can find the failure rate function:

$$h(t) = \frac{2te^{-t^2}}{e^{-t^2}} = 2t.$$

Hence this distribution has increasing failure rate. As a second example, consider the exponential distribution with parameter $\lambda = 5$. By formulas (13) and (15) this distribution has failure rate

$$h(t) = \frac{5e^{-5t}}{e^{-5t}} = 5.$$

Hence the exponential distribution has constant failure rate. Devices with this lifetime distribution do not age in the sense of having a higher likelihood of impending failure in a short interval when they are older than when they are younger.

Exercise

7. Show that, regardless of the value of the parameter λ, the Weibull distribution has increasing failure rate if $\beta > 1$, constant failure rate if $\beta = 1$, and decreasing failure rate if $\beta < 1$.

The failure rate and the result of Exercise 7 allow us to distinguish better among members of the Weibull family. Figure 6 shows sketches of the examples considered in Figure 4, with $\lambda = 2$, and values of $\beta = .5, 1, 1.5$. One sees clearly that $\beta = 1$ (the case of the exponential distribution) is a boundary between the cases of increasing and decreasing failure rate.

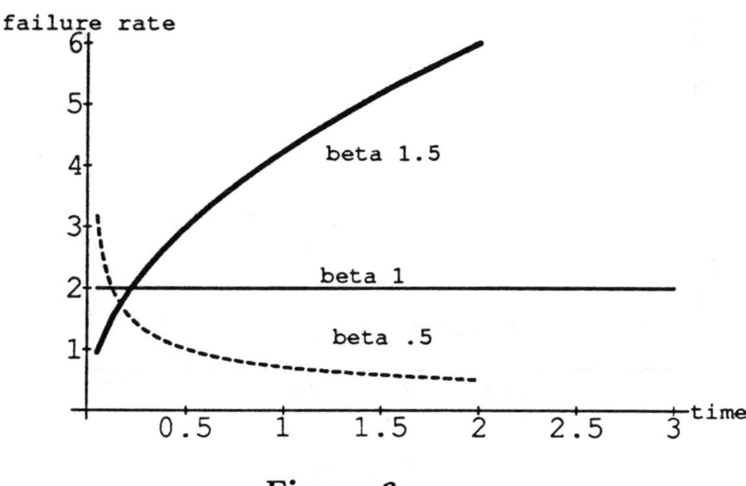

Figure 6

It is possible to check the increasing failure rate condition empirically from a data set of failure times. In the data set of tire lifetimes, we have broken the interval in which data lies into subintervals of size $\Delta t = 5$ thousand. We refer to the points of the partition as $t_0, t_1, t_2, \ldots$, which have the values 20, 25, 30, ... Consider a general interval $[t_i, t_{i+1}]$, where $t_{i+1} = t_i + \Delta t$. Recall that $h(t) = f(t)/(1-F(t))$ and $f(t) = F'(t)$. If we approximate $f(t)$ by the difference quotient of F at t with increment Δt, we obtain

$$h(t) \approx \frac{1}{\Delta t} \frac{F(t+\Delta t) - F(t)}{1 - F(t)}. \tag{25}$$

Since Δt is just a positive constant, to judge whether h is an increasing function, we can plot the following as a function of t_i:

$$\frac{F_N(t_i + \Delta t) - F_N(t_i)}{1 - F_N(t_i)} = \frac{F_N(t_{i+1}) - F_N(t_i)}{1 - F_N(t_i)}, \tag{26}$$

where F_N is the empirical distribution function. But notice that this ratio is just the proportion of tires surviving through time t_i which failed in the next $\Delta t = 5$ time units (which, if divided by Δt, is an intuitive approximation of the failure rate function). The function given in (26) is plotted for our data in Figure 7. For instance, the value of the function at 20 is 2/25, which is the ratio of the number of failures in $[20, 25]$ to the number surviving through time 20. The randomness inevitably results in occasional decreases, but Figure 7 suggests a general upward trend, so we have some empirical support to claim that the distribution of tire lifetimes has increasing failure rate.

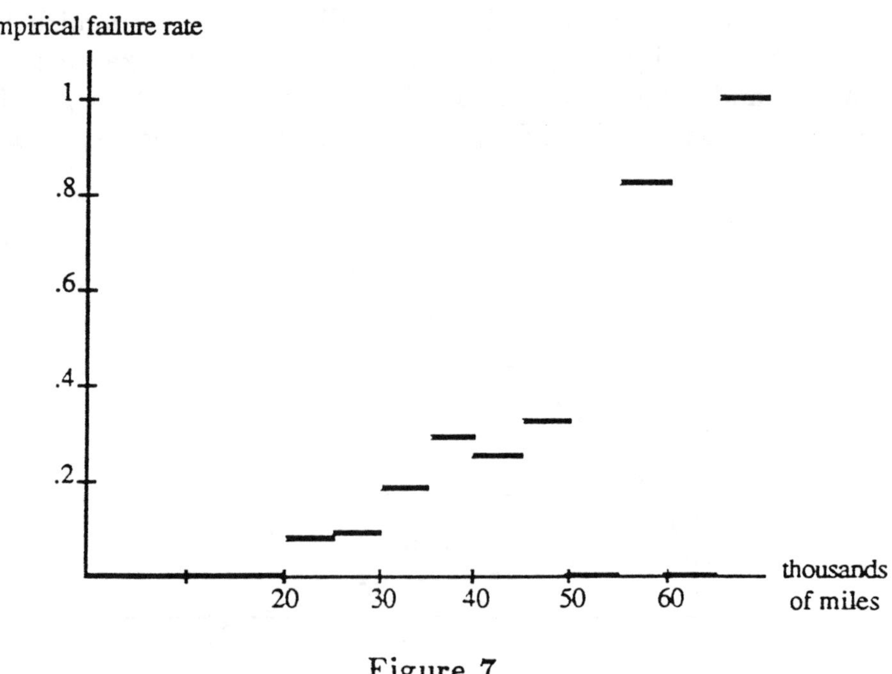

Figure 7

Solution of Tire Problem

The tire life data indicate that the distribution of lifetimes has increasing failure rate. We know a family of distributions, namely the Weibull family with $\beta > 1$, which has increasing failure rate and is frequently a good model for life distributions. Moreover, the empirical reliability function of the tire data (Figure 5) is a good fit to reliability functions of the Weibull family. Hence it is reasonable to suppose that tire lifetimes follow a Weibull distribution.

To specify the probabilistic model, we need only find the parameters λ and β. In the field of statistics, parameter estimation is one of the major problems, but we have no time for a full development here. For a discussion of this particular estimation problem, see (Devore, 1987, p. 243). For our purposes, let us say that a friendly genie familiar with statistics has told us that the parameters are $\lambda = .02$ and $\beta = 4$. (These are in fact the parameters used by the author to simulate the data, chosen to have an average life of around 45 thousand, with modest variability.)

Consider again equation (2) and the "average value of $1 - T/50$ when $T < 50$." In light of our new knowledge about probability distributions, what could that mean? We are averaging the continuum of possible values of $1 - T/50$, and as you have probably seen in your study of calculus, a continuous average is taken to be a limit of discrete averages, which gives rise to a definite integral. We must give higher weight in our averaging process to values which are more likely. If we imagine dividing the interval of interest $[0, 50]$ into many equally sized subintervals $[t_0, t_1], [t_1, t_2], \ldots, [t_{n-1}, t_n]$ of size Δt, then if T takes its value in the ith subinterval $[t_{i-1}, t_i]$ the value of $1 - T/50$ is roughly $1 - t_{i-1}/50$. This event happens with likelihood $F(t_i) - F(t_{i-1}) \approx f(t_{i-1})\Delta t$, by formula (11). So an approximate average of $1 - T/50$ using likelihoods of subintervals as weights is

$$\sum_{i=1}^{n} \left(1 - \frac{t_{i-1}}{50}\right) f(t_{i-1})\Delta t. \tag{27}$$

As the partition becomes finer, this sum converges to a continuous version of the average, which is the first definite integral below.

$$\begin{aligned} \text{average of } 1 - \frac{T}{50} \text{ on } [0,50] &= \int_0^{50} (1 - \frac{t}{50})\ f(t)dt \\ &= \int_0^{50} f(t)dt - \frac{1}{50} \int_0^{50} t\ f(t)\ dt \\ &= F(50) - \frac{1}{50} \int_0^{50} t\ f(t)\ dt. \end{aligned} \tag{28}$$

Using formulas (14) and (16) for the Weibull c.d.f. and density gives the expression

$$\text{avg. of } 1 - \frac{T}{50} \text{ on } [0,50] = 1 - e^{-((.02)(50))^4} - \frac{1}{50} \int_0^{50} t\ (.02)^4 4t^3 e^{-((.02)t)^4}\ dt. \tag{29}$$

Some simple manipulations can be done. By first employing the substitution $s = .02t$, and then using integration by parts with $u = s$ and $dv = 4s^3e^{-s^4}ds$, the preceding expression simplifies to

$$\text{avg. of } 1 - \frac{T}{50} \text{ on } [0, 50] = 1 - \int_0^1 e^{-s^4}\, ds. \tag{30}$$

The last integral, which we will call I, is not expressible in closed form, but it can be approximated using numerical methods like the trapezoidal rule or Simpson's rule. One finds that $I \approx .8448$. Hence, from equation (2), to offset risk of loss, the warranty surcharge w must satisfy

$$p_1 = (p_1 + w)(1 - (1 - I)), \qquad \text{so that} \qquad w = p_1\frac{1-I}{I}. \tag{31}$$

If the tire life distribution that we have been using is the true one, the fact that the total price was $p_1 + w = 75$ implies that $[I/(1-I)]w + w = [1/(1-I)]w = 75$, and therefore I paid $w = (1-I)75 = \$11.64$ for the warranty.

Exercises

8. Derive equation (30) from equation (29).

9. Use Simpson's rule with $n = 6$ to verify that

$$\int_0^1 e^{-s^4} ds \approx .8448.$$

Further Reading

(Gertsbakh, 1989) is a well-written introduction to the subject of reliability. In addition to the material on life distributions which we have extracted for this note, there is much interesting material on replacement and inspection schemes, designed to improve reliability. The reader is advised to study calculus-based probability thoroughly before attempting these things, however. Some references are (Hogg and Tanis, 1988), (Hoel et al, 1971), and at a higher level (Parzen, 1960). Another interesting area of study is structural reliability, which studies how the reliability of a system depends on the organization of the components from which the system is built. A good discussion of structural reliability can be found in (Kaufmann et al, 1977). To comprehend this the reader only needs a small amount of elementary probability, including the idea of independence, and perhaps some experience in discrete mathematics. The veritable bible of reliability theory is (Barlow and Proschan, 1975). Several papers which discuss applicability of the life distributions that we have discussed are (Davis, 1952), (Epstein, 1958), (Harter and Moore, 1976) and (Weibull, 1951). The last is widely recognized as a seminal work in the area.

References

Barlow, R.E. and F. Proschan (1975), *Statistical Theory of Reliability and Life Testing*, Holt, Rinehart and Winston.

Davis, D.J. (1952), "Analysis of Failure Data," *Journal of the American Statistical Association* 47: 113-150.

Devore, J. (1987) *Probability and Statistics for Engineering and the Sciences*, Brooks/Cole.

Epstein, B. (1958), "Exponential Distribution and its Role in Life Testing," *Industrial Quality Control* 15: 2-7.

Gertsbakh, I.B. (1989), *Statistical Reliability Theory*, Marcel Dekker.

Harter, H.L. and A.H. Moore (1976), "An Evaluation of Exponential and Weibull Test Plans," *IEEE Transactions in Reliability* 2: 100-104.

Hoel, P.G, S.C. Port and C.J. Stone (1971), *Introduction to Probability Theory*, Houghton Mifflin.

Hogg, R.V. and E.A. Tanis (1988), *Probability and Statistical Inference*, Macmillan.

Kaufmann, A., D. Grouchko and R. Cruon (1977), *Mathematical Models for the Study of the Reliability of Systems*, Academic Press.

Parzen, E. (1960), *Modern Probability Theory and its Applications*, Wiley.

Weibull, W. (1951), "A Statistical Distribution of Wide Applicability," *Journal of Applied Mechanics* 18: 293-297.

Answers to Exercises

1. $F(t) = \int_0^t \lambda e^{-\lambda s} ds = e^{-\lambda s}\Big|_0^t$

2. $f(t) = F'(t) = -e^{-(\lambda t)^\beta}(-\beta(\lambda t)^{\beta-1}\lambda)$. Probabilities found by integrating f are not changed if the value of f is changed at a point.

3. $P[T > 1000] = 1 - F(1000) = e^{-(.001)(1000)} = e^{-1}$. Alternatively, $P[T > 1000] = \int_{1000}^{\infty}(.001)e^{-(.001)t}dt = -e^{-(.001)t}\Big|_{1000}^{\infty} = e^{-1} \approx .3679$.

4. Use integration by parts with $u = 2t, dv = 2e^{-2t}dt$.

$$\begin{aligned} P[1 \le T \le 5] &= F(5) - F(1) \\ &= 1 - e^{-10} - 10e^{-10} - (1 - e^{-2} - 2e^{-2}) \\ &= 3e^{-2} - 11e^{-10} \approx .4055. \end{aligned}$$

6.
$$P[T > 1000|T > 500] = \frac{P[T > 1000]}{P[T > 500]} = \frac{e^{-1000\lambda}}{e^{-500\lambda}} = \frac{e^{-1}}{e^{-.5}} \approx .6065.$$

7. $h(t) = \beta\lambda^\beta t^{\beta-1}, \quad h'(t) = \beta(\beta-1)\lambda^\beta t^{\beta-2}$.

9. Simpson's rule with $n = 6$ gives $I \approx .8449$.

QUEUEING SYSTEMS

Author: Kevin J. Hastings, Knox College, Galesburg, IL 61401.

Area of application: business, manufacturing, computer science

Calculus needed: limits, definition of derivative, optimization, differential equations, geometric series, L'Hopital's rule.

Related mathematics: probability, differential equations, recurrence relations.

The Problem: Maximizing Profit for Drive-Up Windows at Burger King

The Burger King corporation has been a leader in the restaurant industry in the use of mathematical models to plan production and service systems. The models which they have built have led to such changes as a redesign of the sandwich preparation board to be perpendicular to the counter, and a move to self-service drinks, in both cases to reduce customer service time at the counter, to speed the customer flow, and hence to increase profit (Swart and Donno, 1981). The model that we will discuss is a modified version of a study undertaken in the late 1970's.

Like most fast food franchises, Burger King operates drive-up windows at all of its restaurants. According to the Swart and Donno article, the average service time at the drive-up window in the 1970's was about 45 seconds. It was normal procedure for one clerk to receive the order, fill the order, and check out the customer. A typical restaurant would be limited to at most 80 cars per hour. An average purchase amount at that time was \$2.44, which produced a maximum possible revenue of (\$2.44)(80) = \$195 per hour.

The Burger King corporation set a goal of reducing the average service time to 30 seconds. If this goal could be reached, about 120 cars could be served per hour, producing a maximum revenue of about (\$2.44)(120) = \$292 per hour. Even if only half of the extra capacity were actually realized, this translates to a \$35,000 increase in yearly revenue capacity per restaurant, and potentially \$52,000,000 in increased sales for the corporation. The importance of reducing service time is plain to see. The goal of 30-second average service times was eventually achieved by dividing labor: the tasks of order-taking, order-assembly, and check-out were given to three different individuals. Note, however, that this analysis has not included the cost of hiring more or better employees to give the extra service, nor has it explicitly considered customers lost due to limited waiting space. Taking these things into account, was the Burger King goal a wise one? We will build a model answer this question.

First we need to specify some assumptions and introduce some notation. Suppose that a particular restaurant has a parking area which can accommodate at most $K = 10$ cars in the drive-up lane (see Figure 1). Cars wanting to enter the lane arrive in a random way, at a rate of 1.5 per minute on the average, but when the lane is full they drive away and are lost as customers. Once in line they are served on a first-come, first-served basis. Service time also varies randomly, with average service rate of μ cars per minute. Assume that the average bill per car is \$3.

Also, suppose that every time we speed up service to serve one more car per minute, we incur an extra cost of \$.25 per car, from needing to hire more or better workers. Thus, increasing the service rate should permit more cars to enter the drive-up lane, yielding more revenue, but unfortunately it also increases costs. The question becomes: what value of the service rate μ maximizes the average profit per minute?

The main problem we face is to determine how varying the service rate will affect the number of cars entering the drive-up lane. Since cars which arrive when the lane is full will not enter, we want to compute the probability of having 10 cars in the drive-up lane at any particular time, as a function of the service rate.

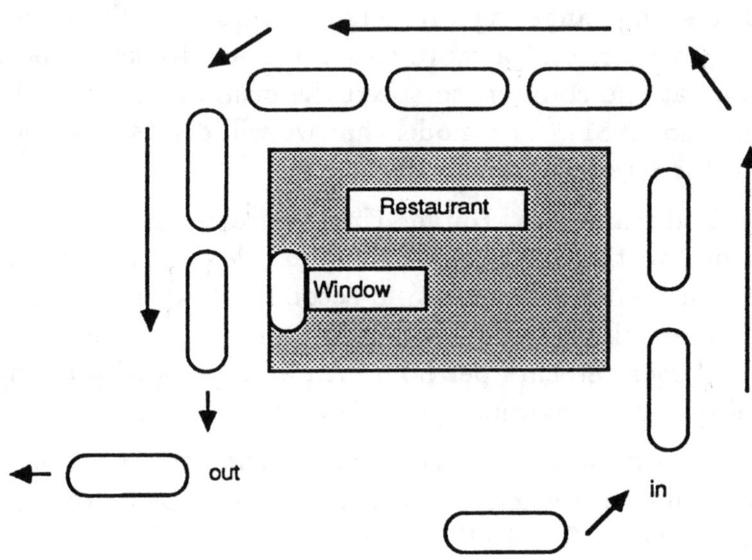

Figure 1

Queueing Theory

A *queue* is a waiting line, and *queueing theory*, the study of waiting lines and other characteristics of service systems, has been one of the most active applications of mathematics in the last several decades. In a queueing system, there are one or more servers, who give service to arriving customers. The customers come at random times, and service of a customer takes a random amount of time. A customer who seeks service from a server who is already occupied must wait in a queue. On completion of service, the customer either proceeds to another service station in the system, or leaves the system.

In our application, we have a very simple configuration of one service station (the drive-up window). Customers (cars) are served on a first-come, first-served basis, and they leave the system when their service is complete. There is a finite amount of waiting space for customers in the queue. We wish to understand how the length of the queue varies with time, since we need to find out how likely it is that the queue for the drive-up window will be full.

The queue length at any time t could be calculated if we knew (1) exactly when each customer would arrive, and (2) exactly how long it would take to serve each customer. Unfortunately, in a real application one cannot expect these times to be known before the actual run of the queueing process. For example, if customers are airplanes arriving at a busy airport like Chicago O'Hare, even though they may be scheduled to land at certain times, random fluctuations invariably occur (mostly in the direction of lateness rather than otherwise). We must assume that arrival times and service times are random variables, whose probability laws are known or can be estimated from data.

One probabilistic assumption which is often made is that there is a number λ, called the *arrival rate*, such that the probability that a new customer will arrive in any short time interval of length Δt is $\lambda \cdot \Delta t$.‡ The corresponding assumption for service times is that there is a number μ, called the *service rate*, such that if a customer is being served at the beginning of a short time interval of length Δt, the probability that the service will be completed within the interval is $\mu \cdot \Delta t$. If these assumptions hold, it can be shown that λ is the average number of customers arriving per unit time, and μ is the average number of customers served per unit time when the server is busy. Hence the names "arrival rate" and "service rate" are appropriate. We will assume that λ and μ exist with these properties.

To solve the Burger King problem, we need to understand the probability distribution of the queue length. Accordingly, we will define, for $n = 0, 1, 2, \ldots$,

$$P_n(t) \equiv \text{the probability that there are } n \text{ customers in the system at time } t. \qquad (1)$$

We will try to calculate the functions $P_n(t)$.

Formulation of Differential Equations; Long-Run Probabilities

Calculus enters the problem because our assumptions about arrival times and service times give us information about how the queue length changes in time. We will use this information to set up *differential equations* for the functions $P_n(t)$. It will be hard to solve the equations explicitly, but we will compute the limiting value of $P_n(t)$ as $t \to \infty$, i.e. the *long-run* probability of n customers being in the system. Once we have finished these computations, we will solve a one-variable optimization problem by the methods of differential calculus.

‡ The assumption is actually that the probability is approximately $\lambda \cdot \Delta t$, and that the approximation approaches equality as $\Delta t \to 0$. (For the experts, we are assuming that the inter-arrival times have an exponential distribution.)

We start by calculating $P_n(t+\Delta t)$, the probability that there are n customers in system at time $t+\Delta t$, in preparation for writing an equation for $P_n'(t)$. Our approach is to envision how the system could have n customers at time $t+\Delta t$, in light of what the situation was a short time earlier at time t. First consider the case $n=0$, i.e. think about the problem of finding the probability of no customers in the system at time $t+\Delta t$. There are three cases:

(a) At time t there were no customers in the system, and no one came during the period $(t, t+\Delta t]$;

(b) At time t there was one customer, who was served during time interval $(t, t+\Delta t]$, and no other customers arrived in that period;

(c) There were two or more arrivals or departures during $(t, t+\Delta t]$, resulting in no customers at time $t+\Delta t$.

If Δt is small, events of type (c) can be shown, under our assumptions, to have probability on the order of $(\Delta t)^2$. Later we will divide the probability of events by Δt, and then let Δt approach zero; hence these terms from case (c) will not contribute to the eventual expression for $P_0'(t)$. Thus, we will ignore them completely and pretend that there are only two simple ways, namely (a) and (b), of having no customers in the system at time $t+\Delta t$. The tree in Figure 2 illustrates the possible situations.

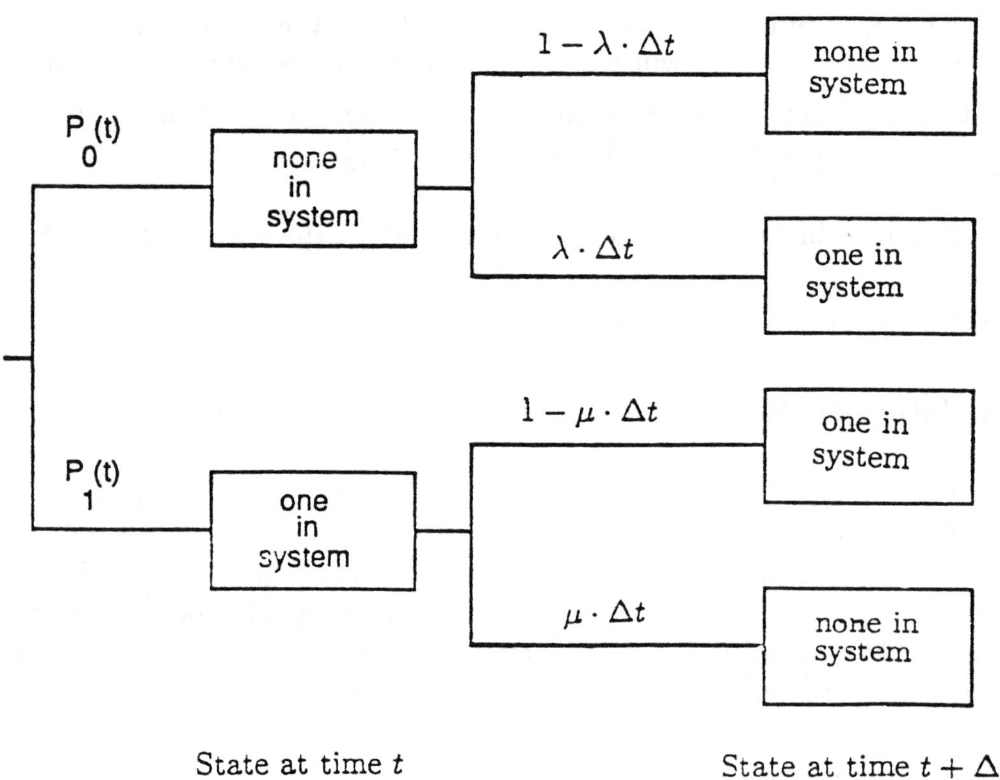

Figure 2

By definition, the quantities $P_0(t)$ and $P_1(t)$ respectively are the probabilities of having 0 or 1 customer in the system at time t. Two of the paths in the tree lead to a system size of 0 at $t + \Delta t$. The total probability $P_0(t + \Delta t)$ is the sum of the probabilities of the paths which lead to this state. Since each path requires each of two steps to occur, its probability is the product of the probabilities of the steps on the path. Hence the tree in Figure 2 gives the approximate equation

$$P_0(t + \Delta t) \approx P_0(t)\ (1 - \lambda \cdot \Delta t) + P_1(t)\ \mu \cdot \Delta t. \tag{2}$$

The error in approximation on the right involves only terms of order $(\Delta t)^2$. Subtracting $P_0(t)$ from both sides of the above yields

$$P_0(t + \Delta t) - P_0(t) \approx -\lambda P_0(t)\Delta t + \mu P_1(t)\Delta t, \quad \text{so}$$

$$\frac{P_0(t + \Delta t) - P_0(t)}{\Delta t} \approx -\lambda P_0(t) + \mu P_1(t)$$

In the limit as $\Delta t \to 0$, the neglected terms on the right side vanish, the approximation becomes equality, and we obtain the differential equation

$$P_0'(t) = -\lambda\ P_0(t) + \mu\ P_1(t). \tag{3}$$

Next, we can get differential equations for $P_n(t), 1 \le n < K$ in a similar way (recall that K is the capacity of the system). First, we argue as in the $n = 0$ case to derive

$$P_n(t + \Delta t) \approx P_{n-1}(t)\ \lambda \cdot \Delta t + P_n(t)(1 - \lambda \cdot \Delta t)(1 - \mu \cdot \Delta t) + P_{n+1}(t)\ \mu \cdot \Delta t. \tag{4}$$

The $\lambda\mu(\Delta t)^2$ term arising from the second term in the sum may be safely ignored. Rearranging terms to find an expression for the difference quotient, and passing to the limit, we get the equation

$$P_n'(t) = \lambda P_{n-1}(t) - (\lambda + \mu)P_n(t) + \mu P_{n+1}(t), \qquad 1 \le n < K. \tag{5}$$

Exercise

1. Give a plausibility argument for equation (4) above, and do the algebraic details leading to equation (5). (Hint: The only non-negligible events permit the system size at time t to be $n-1$, n, or $n+1$. What must happen, given each of these, in order to have n customers in the system at time $t + \Delta t$?)

Finally, in the case $n = K$ it is not possible to have $K + 1$ customers in the system, so that the rightmost term of (4) does not appear, and arrivals are forbidden when the system size is K, so that in the middle term of (4) the quantity $(1 - \lambda\ \Delta t)$ is replaced by 1. The resulting approximation is therefore

$$P_K(t + \Delta t) \approx P_{K-1}(t)\ \lambda \cdot \Delta t + P_K(t)(1 - \mu \cdot \Delta t), \tag{6}$$

which leads to the differential equation

$$P'_K(t) = \lambda P_{K-1}(t) - \mu P_K(t). \tag{7}$$

Equations (3),(5) and (7) together form a system of first order differential equations for the unknown functions $P_n(t), 0 \le n \le K$. They can be solved (see Saaty, 1961, Chapter 5), but it is difficult, and it requires some sophisticated linear algebra and analysis. We will not treat this here, but we can investigate the limiting behavior of $P_n(t)$ as $t \to \infty$. For the original application, this means that we will assume that the system has run for a long enough time to have approached an equilibrium condition, where the probability distribution of the queue length no longer varies in time.

Denote the long-term probability that n customers are in the system by π_n:

$$\pi_n = \lim_{t \to \infty} P_n(t). \tag{8}$$

We take for granted the (true) fact that this limit exists and hence that $P'_n(t)$ approaches 0 as $t \to \infty$. Passing to the limit in equations (3), (5) and (7) then produces the equations

$$0 = -\lambda \pi_0 + \mu \pi_1 \tag{9}$$

$$0 = \lambda \pi_{n-1} - (\lambda + \mu)\pi_n + \mu \pi_{n+1}, \qquad 1 \le n < K \tag{10}$$

$$0 = \lambda \pi_{K-1} - \mu \pi_K, \tag{11}$$

which imply

$$\mu \pi_1 = \lambda \pi_0 \tag{12}$$

$$\mu \pi_{n+1} = (\lambda + \mu)\pi_n - \lambda \pi_{n-1}, \qquad 1 \le n < K \tag{13}$$

$$\mu \pi_K = \lambda \pi_{K-1}. \tag{14}$$

Exercise

2. Use equations (12)–(14) above to show that when $K \ge 3$:

$$\pi_1 = \frac{\lambda}{\mu}\ \pi_0, \qquad \pi_2 = \left(\frac{\lambda}{\mu}\right)^2 \pi_0, \qquad \pi_3 = \left(\frac{\lambda}{\mu}\right)^3 \pi_0.$$

3. Use mathematical induction to prove that

$$\pi_n = \left(\frac{\lambda}{\mu}\right)^n \pi_0, \qquad 1 \le n \le K, \tag{15}$$

The sum of all of the probabilities π_n for values of n from 0 to K must be 1, since that sum represents the probability that in the long run the system size takes on at least one of its possible values. Using this together with the result of Exercise 3 gives

$$\begin{aligned}
1 &= \pi_0 + \sum_{n=1}^{K} \left(\frac{\lambda}{\mu}\right)^n \pi_0 \\
&= \sum_{n=0}^{K} \left(\frac{\lambda}{\mu}\right)^n \pi_0 \\
&= \frac{1-(\frac{\lambda}{\mu})^{K+1}}{1-\frac{\lambda}{\mu}}\,\pi_0, \text{ if } \frac{\lambda}{\mu} \neq 1 \\
&= \frac{1-\rho^{K+1}}{1-\rho}\,\pi_0,
\end{aligned} \tag{16}$$

where we use ρ to denote the quotient λ/μ. The quantity ρ is thus a measure of the rate of arrivals compared to the service rate. It is called the *traffic intensity.* In the third line of the calculation we have used the formula for the finite sum of a geometric series. Equation (16) implies that $\pi_0 = (1-\rho)/(1-\rho^{K+1})$. Combining this with equation (15) gives our final result for the long-run distribution of the number of customers in the system:

$$\pi_n = \frac{1-\rho}{1-\rho^{K+1}}\,\rho^n, \qquad 0 \le n \le K. \tag{17}$$

A graph of this distribution is given in Figure 3, for the values $\rho = .8$ and $K = 10$.

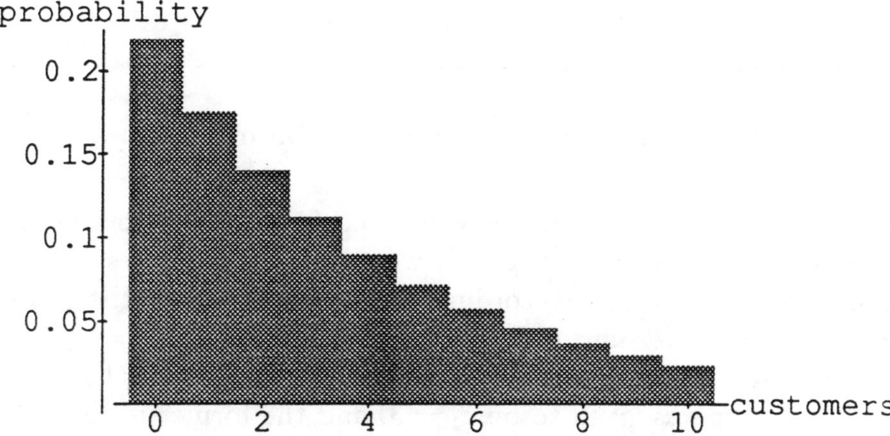

Figure 3

Exercises

4. Show that in the case $\rho = \frac{\lambda}{\mu} = 1$, all of the limiting probabilities π_n must equal $\dfrac{1}{K+1}$.

5. Consider the special case $K = 1$. Then at any given time there are only two possibilities: the number of customers in the system is 1 or 0, according to whether the server is busy or not. Working as in the derivation of this section, show that a system of two differential equations for the quantities

$$\begin{aligned} P_1(t) &= \text{ probability that the server is busy at time } t \\ P_0(t) &= \text{ probability that the server unengaged at time } t \end{aligned}$$

is

$$\begin{aligned} P_1'(t) &= \lambda P_0(t) - \mu P_1(t) \\ P_0'(t) &= -\lambda P_0(t) + \mu P_1(t) \end{aligned}$$

Since $P_0(t)$ and $P_1(t)$ are probabilities of complementary events, $P_0(t) = 1 - P_1(t)$. In light of this, find a single differential equation for the function $P_1(t)$ alone, and show that the following function satisfies the equation:

$$P_1(t) = \frac{\lambda}{\lambda+\mu} - \frac{\lambda}{\lambda+\mu} e^{-(\lambda+\mu)t}.$$

An interesting consequence is that the probability that the server is busy at time t quickly approaches a limiting value of $\lambda/(\lambda+\mu)$ as $t \to \infty$.

6. In the Burger King problem, $\lambda = 3/2$. If the average service time is 45 seconds $= 3/4$ minute, $\mu = 4/3$. For these rates, what is the long-run probability that there will be no customers in the system? (This can be interpreted as the percentage of the time the server will be idle.) Recalculate the probability that there will be no one in the system if the average service time is reduced to 30 seconds, so that $\mu = 2$.

Solution of Burger King Problem

We are to find a service rate μ which maximizes profit per unit time at the drive-up window. Since the arrival rate $\lambda = 1.5$ is given, we may as well express the problem in terms of the traffic intensity $\rho = \lambda/\mu$. According to the conditions of the problem, there is an overhead cost of $c_2 = .25$ per unit of μ, giving a total cost of $c_2\mu = c_2\lambda/\rho$. Revenue is earned at a rate equal to the rate at which customers join the system, times the average bill per customer. The latter was given to be $c_1 = 3$, and the former is the arrival rate λ times the proportion of cars $1 - \pi_K$ which actually enter the drive-up lane. From equation (17), in the case that $\rho \neq 1$, this quantity is

$$1 - \pi_K = 1 - \frac{1-\rho}{1-\rho^{K+1}}\,\rho^K = \frac{1-\rho^K}{1-\rho^{K+1}}. \tag{18}$$

The company's profit to be maximized as a function of ρ is revenue minus cost, i.e.

$$P(\rho) = c_1\lambda\frac{1-\rho^K}{1-\rho^{K+1}} - \frac{c_2\lambda}{\rho}. \tag{19}$$

For our choice of constants, $c_1 = 3, c_2 = .25, K = 10, \lambda = 1.5$ the profit function becomes

$$P(\rho) = 4.5\frac{1-\rho^{10}}{1-\rho^{11}} - \frac{.375}{\rho}, \quad \rho \neq 1. \tag{20}$$

The result of Exercise 4 implies that in the case $\rho = 1$ the profit is

$$\begin{aligned} P(1) &= c_1\lambda(1 - \frac{1}{K+1}) - \frac{c_2\lambda}{1} \\ &= 4.5 \cdot \frac{10}{11} - .375 \approx 3.716. \end{aligned} \tag{21}$$

Figure 4 contains a sketch of this function.

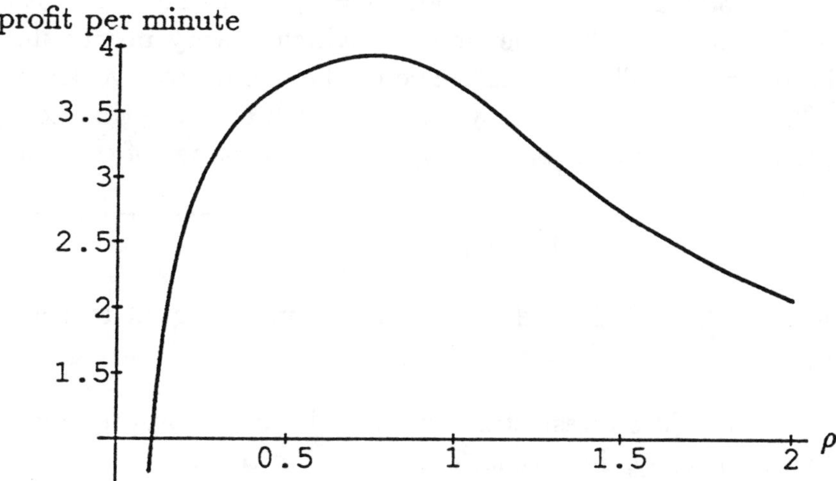

Figure 4

Exercise

7. Show that the profit function P is continuous at $\rho = 1$. (Hint: Use formulas (19) and (21), and apply L'Hopital's rule to the ratio in (19).)

The graph in Figure 4 seems to indicate a maximum at about $\rho = .75$. We can attempt to find the exact value by setting the derivative of P equal to zero. Applying the quotient rule to (19) yields

$$P'(\rho) = \lambda\left[c_1\frac{(1-\rho^{K+1})(-K\rho^{K-1}) - (1-\rho^K)(-(K+1)\rho^K)}{(1-\rho^{K+1})^2} + \frac{c_2}{\rho^2}\right]. \tag{22}$$

The reader should check (Exercise 8) that setting this derivative equal to zero results in the polynomial equation in ρ

$$(c_1 - c_2)\rho^{2K+2} - c_1(K+1)\rho^{K+2} + (c_1K + 2c_2)\rho^{K+1} - c_2 = 0. \tag{23}$$

For our data, the critical point which achieves maximum profit is a solution to the equation

$$2.75\rho^{22} - 33\rho^{12} + 30.5\rho^{11} - .25 = 0. \tag{24}$$

We have encountered a frequent situation in real applications of mathematics: an equation has been derived for the solution, but we do not know how to solve it exactly. However, calculus has by no means failed us, for there are numerical methods (like Newton's method, discussed in most calculus texts) for approximating the desired solution arbitrarily closely. We leave the details to the reader, but applying numerical techniques to find the solution of (24) gives $\rho \approx .750962$. Since $\lambda = 1.5$ and $\rho = \lambda/\mu$, this gives an optimal service rate of $\mu = \lambda/\rho = \lambda/.750962 \approx 1.99744$ per minute, which is very nearly the 30-second goal arrived at by the company. The optimal value of the profit rate works out to about \$3.93 per minute, or \$236.03 per hour. It is by methods such as these that companies like Burger King are able to set operational goals to improve the workings of their organization.

Exercises

8. Simplify expression (22) for $P'(\rho)$ and show that setting it equal to zero yields the polynomial equation (23).

9. Apply Newton's method (I'd suggest starting with the initial approximation .75) or equation-solving software to verify that .750962 is a solution to (24).

10. If you have a graphics device such as Mathematica, MasterGrapher, or a hand-held calculator available, study the sensitivity of the optimal solution of this problem to changes in the parameters c_1, c_2, λ, K. Specifically, as each parameter is allowed to increase while the rest are held fixed, what happens to the maximum profit and to the optimal service rate?

Discussion

Analytical calculations of important characteristics of queues can be difficult, even in a relatively simple queueing system. In fact, the class of problems in the real world that are amenable to exact solution is fairly small, because many queueing systems consist of large networks with complicated physical restrictions and interdependencies, and some of the ingredients of the basic model that we have assumed constant, like the arrival rate, may well vary with time.

However, the situation is not hopeless. When exact solution is impossible, the queueing system can still usually be *simulated* on a computer, and approximate numerical results can be obtained. To simulate, all we need to do is instruct the computer to produce a sequence of random arrival times, and a sequence of random service times, having the statistical characteristics we observe in the system we want to study. This is easier than it sounds, due to the existence of special purpose languages like GPSS, SLAM, and SIMAN for simulation of service systems. Animation packages are now becoming available in which the modeller can actually watch how the system is behaving. An instructional program called GASP (Fisch and Griffeath, 1988) has a module in which you schedule servers at a fast food restaurant to maximize the daily profit. Once you select a schedule, you can watch an animation of the restaurant's operation, and based on this, guess a better schedule.

Our description of queues suggests familiar applications to serving human customers in retail stores and banks, but queueing theory is certainly not limited to this. Planes (customers) waiting to use runways (servers) illustrate one possible application. Recently, American Airlines used queueing theory to help decide whether it was worth a billion dollar investment to build a new terminal on the west side of Dallas-Fort Worth airport, and also to convince regional authorities of the necessity for two new turbojet runways (Cook, 1989). Queueing models are frequently used in manufacturing, specifically in modelling flows of goods (customers) through assembly stations, warehouses, and shipping areas (servers) (Law and McComas, 1988). Perhaps the most active uses of queueing theory today involve telecommunications, in which messages (customers) are switched from station (server) to station, and computer networks, in which jobs (customers) share a network of processors and input-output devices (servers). The second volume of (Kleinrock, 1976) is filled with examples of this sort.

You would need a thorough grounding in probability theory to make a serious study of queueing systems. Two places to start are (Hogg and Tanis, 1988) and (Hoel et al, 1971). A gentle introduction to queueing may be found in (Hillier and Lieberman, 1990). Three nice expositions, at a higher level, of classical queueing theory are (Gross and Harris, 1985), (Kleinrock, 1975) and (Saaty, 1961). New applications and theory appear regularly in the journals *Operations Research* and *Management Science.* (Solomon, 1983) is a good introduction to simulation of queues, and also contains a reprint of the Swart and Donno article on the Burger King studies, as well as other applications.

References

Cook, J.M. (1989), "OR/MS: Alive and Flying at American Airlines," *OR/MS Today* 16: 16-18.

Crabill, T.B., D. Gross and M.J. Magazine (1977), "A Classified Bibliography of Research on Optimal Design and Control of Queues", *Operations Research* 25: 219-232.

Fisch, B, and D. Griffeath (1988), *Graphical Aids for Stochastic Processes*, Wadsworth & Brooks/Cole Advanced Books and Software.

Gross, D. and C. Harris (1985), *Fundamentals of Queueing Theory, 2nd ed.*, Wiley.

Hillier, F.S. and G.J. Lieberman (1990), *Introduction to Operations Research, 5th ed.*, McGraw Hill.

Hoel,P.G., S.C. Port and C.J.Stone (1971), *Introduction to Probability Theory*, Houghton Mifflin.

Hogg, R.V. and E.A. Tanis (1988), *Probability and Statistical Inference*, Macmillan.

Kleinrock, L.M. (1975), *Queueing Systems, Vol.1:Theory*, Wiley.

Kleinrock, L.M. (1976), *Queueing Systems, Vol.2:Computer Applications*, Wiley.

Larson, R.C. (1989), "OR/MS and the Services Industries", *OR/MS Today* 16:12-18.

Law, A.M. and M.G.McComas (1988), "How Simulation Pays Off", *Manufacturing Engineering*, February.

Saaty, T.L. (1961), *Elements of Queueing Theory with Applications*, Dover.

Solomon, S.L. (1983), *Simulation of Waiting Line Systems*, Prentice-Hall.

Swart, W. and L. Donno (1981), "Simulation Modelling Improves Operations, Planning, and Productivity of Fast Food Restaurants," *Interfaces* 11: 35-47.

Answers to Exercises

3. Exercise 2 starts the induction. Assume that the formula holds up to n, and use (13) to calculate

$$\pi_{n+1} = \frac{1}{\mu}\left[(\lambda+\mu)\left(\frac{\lambda}{\mu}\right)^{n}\pi_0 - \lambda\left(\frac{\lambda}{\mu}\right)^{n-1}\pi_0\right] = \left(\frac{\lambda}{\mu}\right)^{n+1}\pi_0.$$

4. Since $\pi_n = (\lambda/\mu)^n\pi_0$ for $1 \leq n \leq K$, in this case it follows that all π_n are equal to π_0. In order for $K+1$ such identical probabilities to sum to 1, each must equal $1/(K+1)$.

6. If $\mu = 4/3$, $\rho = \lambda/\mu = 9/8$. Then $\pi_0 = \dfrac{1-\rho}{1-\rho^{11}} = \dfrac{-.125}{-2.6532} = .047$. If $\mu = 2$, $\rho = 3/4$, and $\pi_0 = \dfrac{.25}{.042235} = .261$. One cost of decreasing the service time is that the percentage of "idle time" of the employees paid to serve increases from about 5% to about 26%.

MOVING A PLANAR ROBOT ARM

Author: Walter Meyer, Adelphi University, Garden City, NY 11530. Adapted by Philip Straffin, Beloit College, Beloit, WI 53511

Area of application: robotics, industry

Calculus needed: derivatives of trigonometric functions, chain rule, parametric representation of motion in the plane. Students also need to solve 2x2 systems of linear equations.

The Problem: Controlling a Robot Arm

Many industrial processes are now carried out by computer-controlled robots. The design and control of robots is the subject of a new intellectual discipline called ***robotics***, which makes heavy use of mathematics, including calculus. In this module we will discuss controlling the motion of a very simple two-dimensional, two-joint robot arm. Figure 1 shows the simplified robot arm we have in mind. There are few, if any, robot mechanisms of this type. However, the concepts we will develop in this simple context are quite similar to the concepts one must wrestle with for more realistic robots such as the one shown in Figure 2.

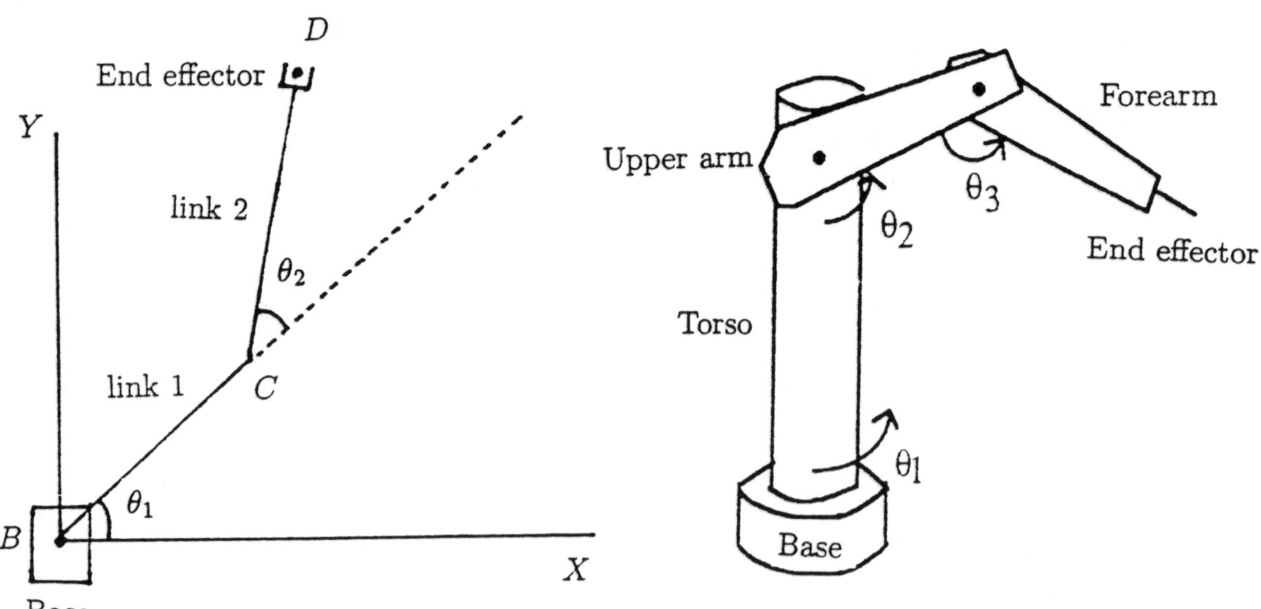

Figure 1. Our Simple Robot.

Figure 2. A More Realistic Robot.

The planar robot has two links, each of which is a line segment. Point B, the base of the robot, is fixed. The first link rotates around point B, and the second link rotates with respect to the first around point C. The entire motion of the robot takes place in the plane. In this plane, we will choose a coordinate system whose origin is at B, and whose positive x-axis points to the right, as in Figure 1. The angle θ_1 between the positive x-axis and link 1, and the angle θ_2 between links 1 and 2, are both controlled by the robot's computer.

The part of the robot which does useful work is at the tip, called the *end effector*. One might imagine a drill or gripper or paint sprayer attached there. Usually, a particular point on the end effector is singled out for attention. In the case of a gripper, this might be the point halfway between the jaws—point D in Figure 1. A key question in robotics is "How can we move the end effector about in its work area?" The problem is that we can't control the end effector directly; instead we must control the angles θ_1 and θ_2 so as to create the desired motion at the end effector. This is a problem for calculus.

The Kinematic Equations

Although we will eventually study the motion of the robot, we need to start with a static problem: given particular values for θ_1 and θ_2, what are the resulting values for x_D and y_D, the coordinates of the end effector point D? This is called the *forward kinematic problem.*

In thinking about this, it is helpful to keep in mind that the angles made by the links are measured in the counterclockwise direction and can be positive or negative. Figure 3 shows some possibilities.

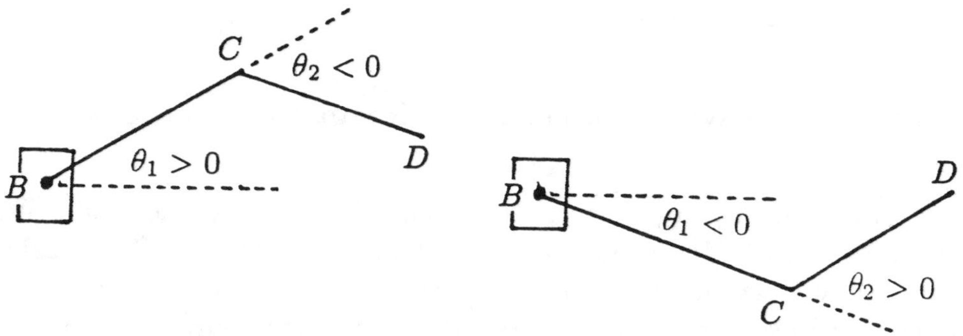

Figure 3. Positive and Negative Angles.

Let us suppose that the lengths of links 1 and 2 are l_1 and l_2 respectively. Figure 4 shows how to derive the kinematic equations. We first find the coordinates x_C and y_C of point C by using the right triangle whose hypotenuse is link 1. Thus

$$x_C = l_1 \cos\theta_1, \qquad y_C = l_1 \sin\theta_1. \tag{1}$$

The coordinates of D are obtained by adding to x_C and y_C the lengths of the appropriate legs of the right triangle whose hypotenuse is link 2:

$$x_D = l_1 \cos\theta_1 + l_2 \cos(\theta_1 + \theta_2), \qquad y_D = l_1 \sin\theta_1 + l_2 \sin(\theta_1 + \theta_2). \tag{2}$$

Although Figure 4 is drawn with θ_1 and θ_2 positive, these equations are true for both positive and negative values of either angle.

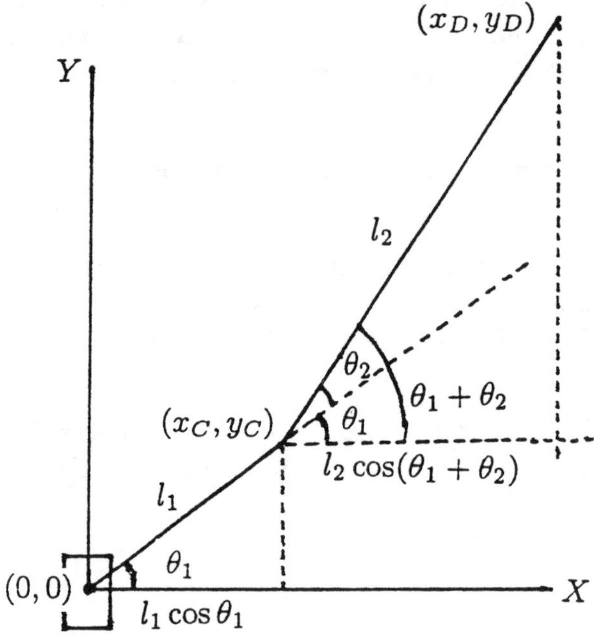

Figure 4. Deriving the Forward Kinematic Equations.

Robots usually have measuring devices to measure the angles at the joints. From these angles, and knowing l_1 and l_2, the robot's computer can use equations (2) to calculate the position of the end effector at any given instant. However, if great accuracy is required this can be unreliable. The measurement of angles is imperfect. It is hard even to know l_1 and l_2 accurately because manufacturing inaccuracies cause them to come out differently from the manufacturer's published specifications. In addition, the lengths of links can change slightly due to temperature fluctuations in the robot's work environment. Despite all this, equations (2), called the *forward kinematic equations*, are among the most heavily used in robotics.

Often we want to turn the forward kinematic thought process around: given a point (x, y) in the robot's work area, what joint angle values θ_1 and θ_2 are required to put the end effector at, or approximately at, (x, y)? In other words, we specify the left hand sides of equations (2) and solve for the angles θ_1 and θ_2. If you try this in Exercise 2, it will challenge your trigonometry and algebra. Knowledge of the resulting *inverse kinematic equations* is not needed to understand this module, although we do need to know that such formulas exist. Note by examining Figure 5 that there are normally two possible solutions, one called *elbow regular* with $\theta_2 > 0$, and the other called *elbow irregular* with $\theta_2 < 0$. We can find formulas to describe each solution.

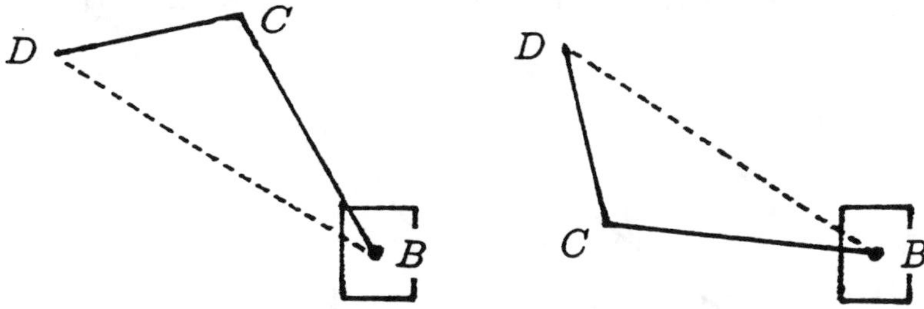

Figure 5. Elbow Regular and Elbow Irregular Inverse Kinematic Solutions.

Exercises

1. Describe the work area, i.e. the region of the plane which can be reached by the end effector,

   a. when $l_1 > l_2$ b. when $l_1 < l_2$.

2. Derive the inverse kinematic equations. Here's a suggested path.

   a. In (2), solve for $\cos(\theta_1 + \theta_2)$ and $\sin(\theta_1 + \theta_2)$, substitute the solutions into $\sin^2(\theta_1 + \theta_2) + \cos^2(\theta_1 + \theta_2) = 1$, and simplify to get

$$y \sin\theta_1 + x \cos\theta_1 = A, \qquad \text{where } A = \frac{x^2 + y^2 + l_1^2 - l_2^2}{2l_1}.$$

   b. Solve this equation along with $\sin^2\theta_1 + \cos^2\theta_1 = 1$ for $\cos\theta_1$, getting

$$\cos\theta_1 = \frac{Ax \pm y\sqrt{x^2 + y^2 - A^2}}{x^2 + y^2}.$$

   Note that because $\cos\theta_1 = \cos(-\theta_1)$ this usually gives four possible values for θ_1. One corresponds to the elbow regular position, one to the elbow irregular position, and two are extraneous roots. Once θ_1 is determined, it is easy to find θ_2 using equations (2).

Velocity Control

Suppose we have a point $T = (x, y)$ in mind and we wish the end effector to pass through it with certain horizontal and vertical velocities. The way to think about this mathematically is to imagine the end effector moving along a curve parameterized by time t. In other words, the curve γ is specified by two functions $x(t)$ and $y(t)$, which give the coordinates of the end effector at time t. A familiar result from calculus is that the velocity vector of the end effector at time t is then the tangent vector $\gamma' = (x'(t), y'(t))$, as shown in Figure 6.

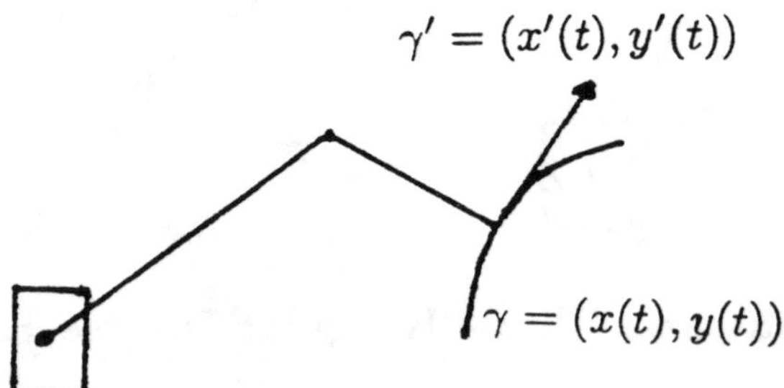

Figure 6. Velocity of the End Effector.

For example, suppose we are washing a vertical window (Figure 7a). We would like there to be no motion in the x direction, i.e. $x'(t) = 0$. We want to be sure the vertical velocity is not too fast, so the window really gets clean, and not too slow so the job doesn't take longer than it has to. For the sake of the example, suppose we want to be moving upward at one foot per second, i.e. $y'(t) = 1$. If we wanted to move along a tilted window, as in Figure 7b, we would need non-zero velocities in both the x and y directions.

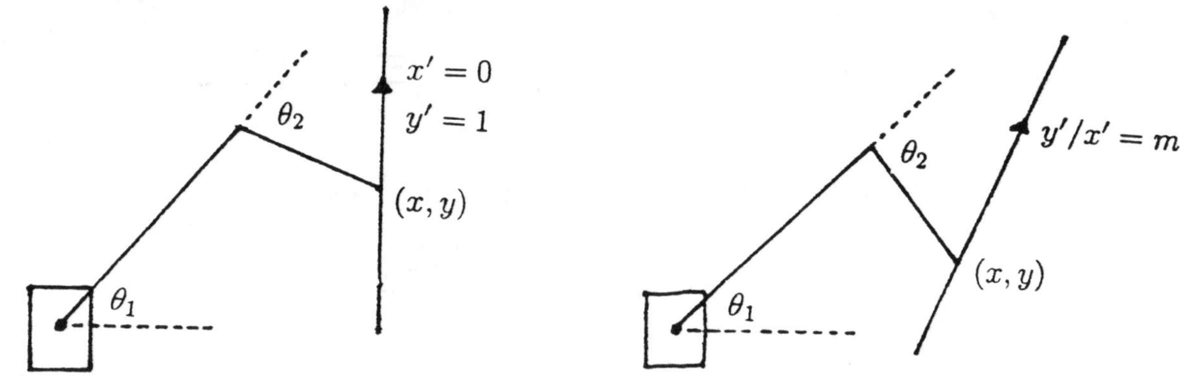

a. Washing a Vertical Window **b. Washing Tilted Surface**

Figure 7.

With x, y, θ_1 and θ_2 written as functions of time, the forward kinematic equations are

$$\begin{aligned} x(t) &= l_1 \cos(\theta_1(t)) + l_2 \cos(\theta_1(t) + \theta_2(t)) \\ y(t) &= l_1 \sin(\theta_1(t)) + l_2 \sin(\theta_1(t) + \theta_2(t)). \end{aligned} \tag{3}$$

Using the chain rule, we can differentiate each of these equations with respect to t:

$$\begin{aligned} x'(t) &= -l_1 \sin(\theta_1(t))\theta_1'(t) - l_2 \sin(\theta_1(t) + \theta_2(t))\ [\theta_1'(t) + \theta_2'(t)] \\ y'(t) &= \ \ l_1 \cos(\theta_1(t))\theta_1'(t) + l_2 \cos(\theta_1(t) + \theta_2(t))\ [\theta_1'(t) + \theta_2'(t)]. \end{aligned} \tag{4}$$

Doing a little algebra, and suppressing t from the notation for simplicity, we obtain

$$\begin{aligned} x' &= -[l_1 \sin\theta_1 + l_2 \sin(\theta_1 + \theta_2)]\ \theta_1' - [l_2 \sin(\theta_1 + \theta_2)]\ \theta_2' \\ y' &= \ \ [l_1 \cos\theta_1 + l_2 \cos(\theta_1 + \theta_2)]\ \theta_1' + [l_2 \cos(\theta_1 + \theta_2)]\ \theta_2'. \end{aligned} \tag{5}$$

By using (2), this can be further simplified to

$$\begin{aligned} x' &= -y\theta_1' - l_2 \sin(\theta_1 + \theta_2)\ \theta_2' \\ y' &= \ \ x\theta_1' + l_2 \cos(\theta_1 + \theta_2)\ \theta_2'. \end{aligned} \tag{6}$$

In our window-washing example, x' and y' have been chosen to be 0 and 1. The point (x, y) is known, so we can apply the inverse kinematic equations to find θ_1 and θ_2. Hence the only unknowns in (6) are θ_1' and θ_2', and we will be able to solve for them. For example, if

$$l_1 = 3.0 \qquad l_2 = 2.0 \qquad x = 4.0531 \qquad y = 1.6037$$

then the inverse kinematic equations would give (Exercise 3)

$$\theta_1 = 0.7854 \text{ radians (45 degrees)} \qquad \theta_2 = -1.0472 \text{ radians } (-60 \text{ degrees}).$$

(Notice that this is the "elbow irregular" configuration.) Then

$$l_2\ \sin(\theta_1 + \theta_2) = -.5176 \qquad l_2\ \cos(\theta_1 + \theta_2) = 1.9318.$$

Making these substitutions, we obtain the following system of equations for the angle velocities:

$$\begin{aligned} 0 &= -1.6037\ \theta_1' + 0.5176\ \theta_2' \\ 1 &= \ \ 4.0531\ \theta_1' + 1.9318\ \theta_2'. \end{aligned} \tag{7}$$

Solving this systems gives

$$\theta_1' = 0.0996 \text{ radians per second} \qquad \theta_2' = 0.3086 \text{ radians per second.}$$

This tells us how fast the robot joints must be turning at the instant when $\theta_1 = 0.7854$ radians and $\theta_2 = -1.0472$ radians in order to have the right velocity for the end effector.

An interesting feature of the solution is that we have to think about time, and introduce it into the notation, in order to do the analysis, but it is not necessary to specify what value time has at the instant when the end effector passes through (x, y). Nor can we solve for time in the equations.

Let's consider briefly a second example, where we want the end effector to move not vertically, but along a line of slope $m = 8/5$. In this case, the y and x velocities of the end effector must have m as their ratio:

$$\frac{y'}{x'} = \frac{8}{5}.$$

If we want a total speed of 0.8 feet per second, we must have

$$\sqrt{(x')^2 + (y')^2} = 0.8.$$

From these equations we can solve for x' and y', determining $x' = 0.4240$ and $y' = 0.6784$.

Suppose that the point (x, y) is the same as in the previous example, $(4.0531, 1.6037)$. Then the angles will be the same as well, and indeed the whole right side of (7) will be the same. The equations for the angular velocities are

$$\begin{aligned} 0.4240 &= -1.6037\,\theta_1' + 0.5176\,\theta_2' \\ 0.6784 &= 4.0531\,\theta_1' + 1.9318\,\theta_2', \end{aligned} \tag{8}$$

so $\theta_1' = -.0901$ radians per second and $\theta_2' = .5401$ radians per second.

Exercises

3.a. Use the result of Exercise 2 to verify the values of θ_1 and θ_2 for the elbow irregular position in the vertical window-washing example.

b. Use the result of Exercise 2 to find the values of θ_1 and θ_2 for the elbow regular position for the same example. [Drawing a careful picture will help pick out the correct value for θ_1. You should find that it is a little bit negative.] Solve for the values of θ_1' and θ_2' needed to get the motion $x' = 0,\ y' = 1$ in the elbow regular position.

4. In the vertical window-washing example, suppose the robot links move for 1/10 second at the angular velocities in the solution to (7). Use the forward kinematic equations to find the new location of the end effector. Notice that the new x-coordinate should be close to the same value of 4.0531, while the new y-coordinate should be close to $1.6037 + 0.1$. Are they?

5. Suppose, using the same data $l_1 = 3,\ l_2 = 2,\ x = 4.0531,\ y = 1.6037$, that we want the end effector to move leftward along a line of slope $-1/2$, at a speed of 1 foot per second. What values of θ_1' and θ_2' will accomplish this?

Teleoperation

Sometimes robotic mechanisms are used as remote control devices with human beings providing continuous "steering" to control the movement of the end effector. This is called *teleoperation*, because the human controller is often far from the robot and watches what is going on at the worksite on a television monitor whose camera is mounted near the robot

(Figure 8). This would not be very efficient for manufacturing, where the idea is to replace repetitive human labor. However, it is useful in jobs which would be hazardous for humans, such as maintenance in nuclear power plants and tasks under the sea, including repair of oil rigs, exploration, and salvage operations. NASA has proposed using teleoperated robots to help construct the Space Station in the 1990's. In these kinds of applications, the tasks and movements the robot must perform are not repetitive, but are often unexpected and novel, where it is beyond the capability of current robotics software to determine appropriate robot actions and carry them out. This is the reason for "putting a human in the loop."

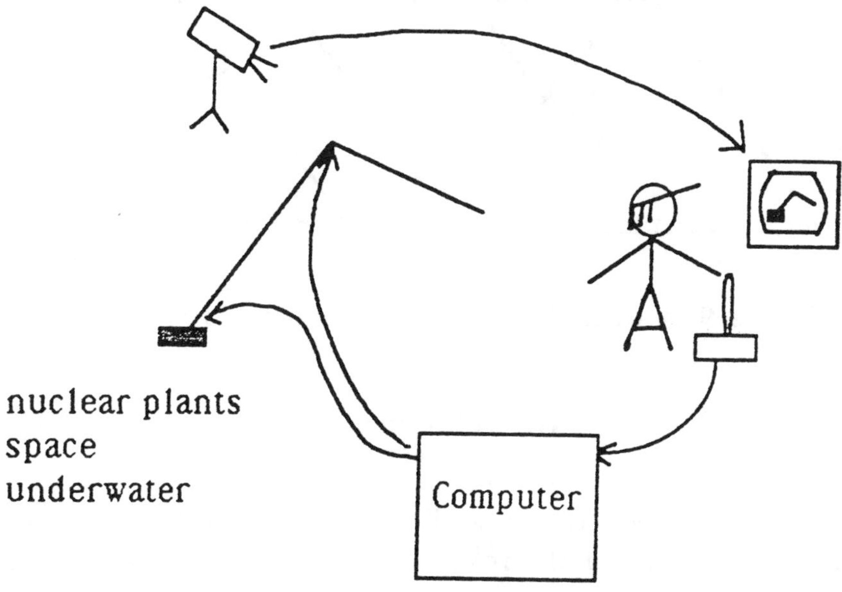

Figure 8. Teleoperation.

The basic strategy in teleoperation is to give the human teleoperator control over the velocity of the end effector (what we have been denoting x' and y' for our planar robot) using a "joystick" or similar device (see Figure 9). Note that the joystick controls velocity, not position. This is similar to the way we drive a car: our foot on the accelerator controls the velocity of the car. Thus, positioning the joystick to the left of center commands a negative value of horizontal velocity ($x' < 0$), the magnitude of the velocity being proportional to how far to the left the joystick is. A deflection of the joystick up or down signals either an upward velocity ($y' > 0$) or a downward velocity ($y' < 0$) of the end effector. Joystick positions combining forward and lateral deflections yield combinations of x' and y'. Leaving the joystick in its "zero" position (pointing straight up) leaves the end effector exactly where it is ($x' = y' = 0$).

The basic scheme of computation carried out by the robot's computer is this. First, the joystick position is read and turned into numerical values for x' and y'. (Figure 9 suggests how this might be done.) Then, using the current values of θ_1 and θ_2 which are read from the robot, the computer gets θ_1' and θ_2' just as we did in the examples above. The motors at the two joints then change the angles at the computed angular velocities for a short time (typically a fraction of a second). Then the new values for the joystick are read and the process starts over.

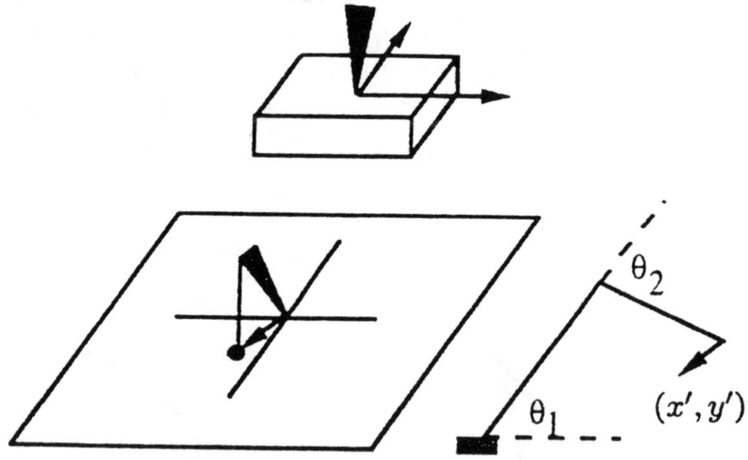

Figure 9. Controlling Velocity with a Joystick.

Singularities

One of the obstacles to teleoperation, and other forms of robot control as well, is the fact that in some configurations of the robot there are some joystick commands which cannot be carried out. This is not because of any shortcoming in the joystick or the computer, but a difficulty inherent in the basic geometric structure of the robot.

Let's look at an example of one of these situations. Suppose the robot is in the configuration shown in Figure 10, with the second link folded completely over onto the first one. Suppose we ask for a motion of the end effector directly away from the point B, say $x' = 0.5$, $y' = 0.5$. This motion appears to be impossible with rigid robot links. If our joystick deflection commands this, something will go wrong somewhere. In order to see what goes wrong, we try to solve for θ_1' and θ_2'. We have

$$l_1 = 3 \qquad l_2 = 2 \qquad x = \frac{1}{\sqrt{2}} \qquad y = \frac{1}{\sqrt{2}} \qquad \theta_1 = \frac{\pi}{4} \qquad \theta_2 = \pi.$$

Substituting into (6) yields

$$\begin{aligned} 0.5 &= -\frac{1}{\sqrt{2}}\,\theta_1' + \sqrt{2}\,\theta_2' \\ 0.5 &= \frac{1}{\sqrt{2}}\,\theta_1' - \sqrt{2}\,\theta_2' \end{aligned} \tag{9}$$

These equations are *inconsistent*: the lefthand sides are the same and non-zero, but the righthand sides are negatives of each other. It is not possible to solve these equations for

θ_1' and θ_2'. No computer code we write, no matter how clever, will be able to find solutions, because there are none to be found.

Indeed, because the righthand sides of equations (9) are negatives of each other, a solution will not exist unless the requested motion has $y' = -x'$. Geometrically, this corresponds to a motion orthogonal to the aligned links of the robot in Figure 10, which we should be able to attain by changing either θ_1 or θ_2 (see Exercise 6). What the equations say is allowed agrees nicely with our geometric intuition.

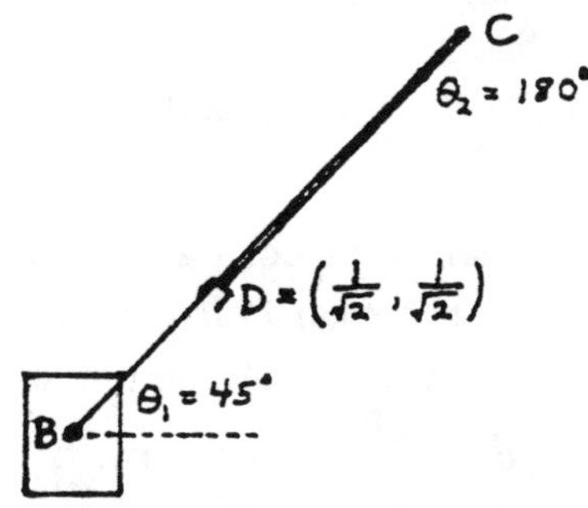

Figure 10. A Singular Configuration.

In general, a robot configuration in which some end effector velocities are impossible to achieve is called a *singular configuration.* Singular configurations are so disruptive that it would be valuable to have a complete understanding of when they occur. In the case of our simple two-link planar robot, this is not difficult. To work out the answer, we will try to find a general solution to equations (6). Multiply the top equation by $\cos(\theta_1+\theta_2)$ and the bottom by $\sin(\theta_1+\theta_2)$ and add to eliminate θ_2'. This yields

$$[x\sin(\theta_1+\theta_2) - y\cos(\theta_1+\theta_2)]\,\theta_1' = [x'\sin(\theta_1+\theta_2) + y'\sin(\theta_1+\theta_2)].$$

The next step would be to solve for θ_1' by dividing by $[x\sin(\theta_1+\theta_2) - y\cos(\theta_1+\theta_2)]$. The only case in which we can't solve for θ_1' occurs when this term is zero. Exactly the same difficulty arises if we try to solve for θ_2'. We conclude

Theorem. *The only singular configurations of the two-link planar robot are those in which* $x\sin(\theta_1+\theta_2) - y\cos(\theta_1+\theta_2) = 0$.

For a geometric interpretation of these singular configurations, see Exercise 7.

Unfortunately, for actual three-dimensional robots such as the one shown in Figure 2, the singularity problem is harder. Given a configuration of the robot, i.e. a specification of all the joint angles, it is possible to determine whether the configuration is singular or not. However, it is not always easy for teleoperators to recognize such configurations by eye. Consequently, they might inadvertently steer the robot into a singular configuration where its mobility is restricted and control becomes unpredictable or breaks down. Gauging the extent of this difficulty and finding ways of overcoming it are important topics in current research. The reference by Baker gives an interesting discussion of the singularity problem.

One common method for alleviating, although not curing, the problem is to add more links to the robot, which designers call "adding redundancy." This is explored in Exercise 8.

Exercises

6. Solve the analogue of equations (9) when $x' = 0.5,\ y' = -0.5$. You should find that the solution is not unique: many different joint rotations will achieve this motion of the end effector. Find a solution with

   a. $\theta_2' = 0$ b. $\theta_1' = 0$ c. $\theta_2' = 0.1$.

7.a. Show that the condition in the singular configuration Theorem is equivalent to saying that $\theta_2 = 0$ or π. Interpret this geometrically. [Notice that the resulting locations of the end effector are all on the boundary of the work area you found in Exercise 1.]

   b. Suppose $x \sin(\theta_1 + \theta_2) - y \cos(\theta_1 + \theta_2)$ is not zero, but is close to zero. This could still be unfortunate. Explain why. In other words, what difficulties could arise if the robot just got close to a singular configuration?

8. Consider a planar three-link robot, based at the origin, with link lengths $l_1 = 4,\ l_2 = 3,\ l_3 = 2$.

   a. Describe the work area of this robot.

   b. Find and draw a singular configuration of this robot in which the end effector is not on the boundary of the work area. In your configuration, which motions of the end effector are not possible? Which motions are possible? [You don't need to work out the kinematic equations. Just reason geometrically.]

   c. Draw a non-singular configuration in which the end effector is at the same location as in your singular configuration.

   d. In general, for a given point (x, y), how many solutions do you think the inverse kinematic equations will have? In other words, how many different robot configurations will put the end effector at (x, y)? [For the two-link robot, there were two.]

9. Building on your answers to Exercise 8, discuss some of the new opportunities and new problems which arise from adding a third link to the planar robot.

References

Baker, Daniel (1990), "Some topological problems in robotics," *The Mathematical Intelligencer* 12: 66-76.

Brady, Michael et al. (1982), *Robot Motion*, MIT Press.

Paul, Richard (1981), *Robot Manipulators*, MIT Press.

Answers to Exercises

1.a. An annulus with outer radius $l_1 + l_2$ and inner radius $l_1 - l_2$.

b. Same except inner radius is $l_2 - l_1$.

3.b. $\theta_1 = -.0319$ radians, $\theta_2 = 1.0472$ radians. The equations are

$$\begin{aligned} 0 &= -1.6037\theta_1' - 1.6993\theta_2' \\ 1 &= \ \ 4.0531\theta_1' + 1.0547\theta_2' \end{aligned}$$

This gives $\theta_1' = .3274$ radians/second, $\theta_2' = -.3101$ radians/second.

4. $\theta_1 = .7854 + .00996 = .7954$; $\theta_2 = -1.0472 + .03086 = -1.0163$. The new x-coordinate is 4.0514, close to 4.0531. The new y-coordinate is 1.7042, close to 1.7037.

5. $x' = -.8944 \qquad y' = .4472 \qquad \theta_1' = .3771 \qquad \theta_2' = -.5596.$

6.a. $\theta_1' = -1/\sqrt{2} = -.7071.$

b. $\theta_2' = 1/2\sqrt{2} = .3536.$

c. $\theta_1' = -.5071.$

7.a. In the condition from the theorem, substitute for x and y using equations (2). Terms cancel, leaving

$$\begin{aligned} 0 &= l_1 \cos\theta_1 \sin(\theta_1 + \theta_2) - l_1 \sin\theta_1 \cos(\theta_1 + \theta_2) \\ &= l_1 \sin(\theta_1 + \theta_2 - \theta_1) \qquad \text{by the angle difference formula} \\ &= l_1 \sin\theta_2. \end{aligned}$$

Thus $\theta_2 = 0$ or π. The second link is either stretched out completely, or folded back over the first link.

b. If we divide by a term close to 0, we may get large angular velocities. Near a singularity, the computer may tell a joint to move faster than it physically can. Hence the motion will not be as expected.

8.a. A disk about the origin of radius 9.

b. Any example where the three links are folded over each other, for instance

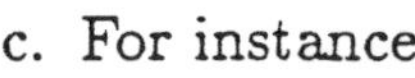

c. For instance

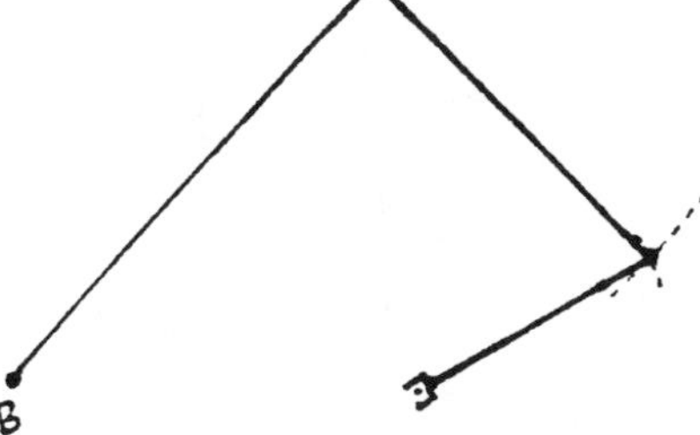

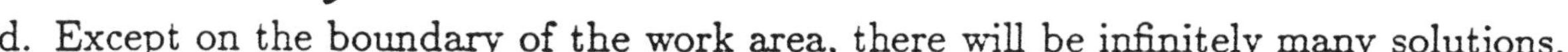

d. Except on the boundary of the work area, there will be infinitely many solutions.

9. An advantage is the freedom to avoid singular configurations. A new problem arises from the indeterminacy of the inverse kinematic equation: from infinitely many possible solutions, we need to pick one. Naturally, we would like to do this in a way which avoids ever moving near singular configurations. There are many interesting mathematical problems here, even for planar robots. For 3-D robots, topology gives some elegant impossibility results—see [Baker, 1990].

DESIGNER CURVES

Author: Steven Janke, Colorado College, Colorado Springs, Colorado 80903

Calculus Needed: derivatives, chain rule. Parametric curves are introduced in the module.

Area of Application: Computer Aided Design

Comments: Software to graph functions and parametric curves is helpful. Software which can use control points to draw Bezier curves would enable students to experiment with the ideas introduced in the module.

The Problem: Designing with a Computer

Before widespread use of computers, design drawings had to be sketched by hand. Any straight lines are easy to draw using a ruler, but the curves are a different matter. An artist can do freehand sketches with sleek, smooth curves, and a draftsperson can use various plastic templates to form curves. But then how do we give the factory exact specifications for such curves?

With powerful new computer aided design systems, it is much more efficient for the designers to work interactively with a computer to visualize and improve their designs. However, the computer needs a careful description of any curves before they can be displayed on the screen. Look at the car profile in Figure 1. The nose of the car looks like a piece of a parabola. If it is, then we simply describe the curve with a quadratic function such as $x = y^2$. But if it is not quite a parabola, perhaps flatter on one side than another, how do we describe it?

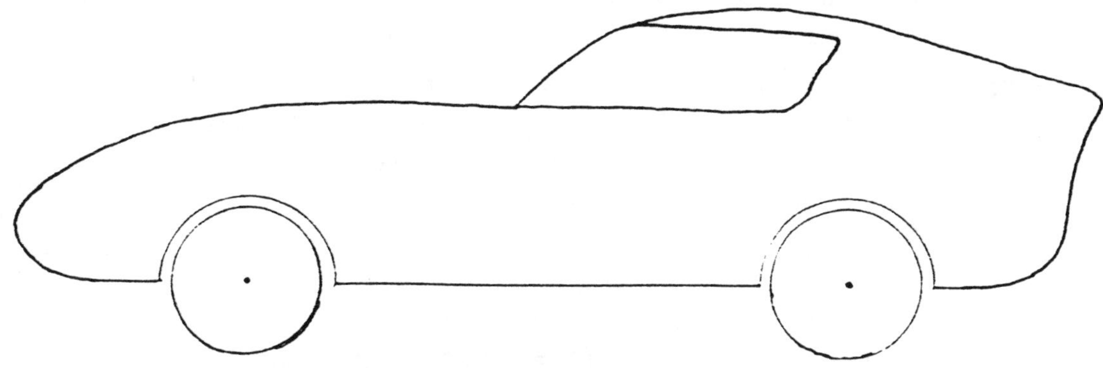

Figure 1.

The problem, then, is to find some system for describing a wide variety of curves. Once we have the description of a particular curve, we need an easy way to alter the curve slightly as we continue to improve the design. If we can mathematically describe what we are doing, then the computer can help in the process and when we are done, there will be nice mathematical descriptions that can control the machines on the factory floor.

Parametric Curves

The most flexible way to represent curves in the plane is by *parametric description.* We introduce a *parameter*, say t, and then give both coordinates x and y as functions of t. Specifying a finite interval for t gives just a piece of the whole curve. For example, the description

$$x = t, \quad y = t^2 \qquad 0 \le t \le 1$$

gives the same curve segment as the description $y = x^2$ for $0 \le x \le 1$. (We can verify this by simply noting that the first equation says $t = x$ and by plugging this into the second equation we get $y = x^2$.) There doesn't seem to be any advantage for the parametric description in this case. However, consider this description:

$$x = 2\cos(t), \quad y = 2\sin(t) \qquad 0 \le t < 2\pi.$$

This parametric description gives the circle $x^2 + y^2 = 4$. (Again we can verify this by noting that $x^2 + y^2 = 4\cos^2 t + 4\sin^2 t = 4$.) The computer can easily find points by looping through the t values and calculating sines and cosines. So the parametric description is nice for computer applications.

Notice that parametric descriptions are not unique. The description

$$x = 2t, \quad y = 4t^2 \qquad 0 \le t \le 1/2$$

gives the same parabolic segment as above. The description

$$x = 2\cos(2\pi - t), \quad y = 2\sin(2\pi - t) \qquad 0 \le t < 2\pi$$

gives the same circle we saw before. Note, however, that the parabolic segment is traced twice as quickly, and the circle is traced in the opposite direction.

Parametric descriptions give direct algorithms for calculating the coordinates of points on the curve, and at the same time that they allow us to describe closed curves and curves that intersect themselves. These two attributes make parametric descriptions particularly useful in computer design systems.

Exercises

1. Show that the equations $x(t) = 2\cos t$, $y(t) = 3\sin t$ describe an ellipse.

2. Graph the curve $x(t) = t(1-t)$, $y(t) = t(1-t^2)$ for $-.1 \le t \le 1.1$. Notice that this curve intersects itself.

3. Show that the description $x(t) = t^2$, $y(t) = t^4$, $0 \le t \le 1$ gives the same curve segment as $x(t) = t$, $y(t) = t^2$, $0 \le t \le 1$ but that the segments are traced out at different rates as t goes from 0 to 1. If the interval were $-1 \le t \le 1$, would the curve segments still be the same?

Combinations of Points

Suppose we have two points $P_0 = (1,3)$ and $P_1 = (3,7)$. It might be necessary in the middle of some design to find a curve from P_0 to P_1. If the curve is a straight line segment, a parametric description is easy to find. Think of t as representing the fraction of the distance from P_0 to P_1, so t runs from 0 to 1. If $t = \frac{1}{4}$, for example, x is one-fourth of the way from 1 to 3, and y is one-fourth of the way from 3 to 7. So $x = 1 + \frac{1}{4}(3-1)$ and $y = 3 + \frac{1}{4}(7-3)$. In general,

$$\begin{aligned} x &= 1 + t(3-1) = t \cdot 3 + (1-t) \cdot 1 \\ y &= 3 + t(7-3) = t \cdot 7 + (1-t) \cdot 3. \end{aligned}$$

We can simplify this parametric description by noting that the expressions for both the x and y coordinates have the same form: $1-t$ times the coordinate from P_0 plus t times the coordinate from P_1. Hence if we let $P(t)$ be a point on the line segment, then

$$P(t) = (1-t) \cdot P_0 + t \cdot P_1 \qquad 0 \le t \le 1.$$

This single equation gives the entire parametric description of the line segment. We say that $P(t)$ is a combination of the points P_0 and P_1. (This combination is sometimes called a "barycentric" combination or a "convex" combination.)

An easy generalization comes to mind. Let $P_0, P_1, \ldots, P_n$ be points in the x-y plane and $\alpha_0(t), \alpha_1(t), \ldots, \alpha_n(t)$ be functions of t. Define $P(t)$ by

$$P(t) = \alpha_0(t)P_0 + \alpha_1(t)P_1 + \cdots + \alpha_n(t)P_n.$$

The point $P(t)$ is a combination of $n+1$ given points. This compact description represents the parametric expressions

$$\begin{aligned} x(t) &= \alpha_0(t)x_0 + \alpha_1(t)x_1 + \cdots + \alpha_n(t)x_n \\ y(t) &= \alpha_0(t)y_0 + \alpha_1(t)y_1 + \cdots + \alpha_n(t)y_n. \end{aligned}$$

Here we have used the notation $x(t)$ and $y(t)$ to emphasize that the coordinates are functions of the parameter t. This method of forming parametric descriptions by taking combinations of points is useful to us since, as we will see, by picking the $\alpha_i(t)$ carefully we can get a variety of curves that either go through or get close to the points P_0, P_1, ... , P_n. The functions $\alpha_i(t)$ are often called "blending" functions since they determine how the given points are blended together to give points on the curve. The points themselves are called "control" points since they control the shape of the curve.

Exercise

4. In the line segment example, $\alpha_0(t) + \alpha_1(t) = 1$. Show that if this property is true, then $P(t) = \alpha_0(t)P_0 + \alpha_1(t)P_1$ lies on a straight line through the points P_0 and P_1. What if $\alpha_0(t) + \alpha_1(t) \neq 1$?

Interpolation

One of the easiest ways of designing a curve is to specify a few control points and then find a curve that goes through those points. We saw in the previous section that if we specify two points then it is easy to describe the straight line that goes through them. Finding curves that go through points is not always easy, but if we concentrate on curves described by polynomials, the problem is tractable.

Recall from algebra that there is a unique straight line that goes through two control points and there is a unique parabola (quadratic curve) of the form $y = ax^2 + bx + c$ that goes through three control points. Be careful here. There can be other parabolas through the three points, but there is a unique one of the given form. In general, if we are given $n+1$ points, we can find an nth degree polynomial that goes through the points. When a curve goes through the control points, we say it *interpolates* those points.

Let's quickly recall how to find these interpolating polynomials. Suppose we wish to find a quadratic curve that goes through the points $P_0 = (-1, 9)$, $P_1 = (2, 3)$, and $P_2 = (3, 5)$. There is such a curve with explicit description $y = ax^2 + bx + c$. The problem is to determine a, b, c. Since the three points are on the curve, we have the following

$$\begin{aligned} 9 &= a - b + c \\ 3 &= 4a + 2b + c \\ 5 &= 9a + 3b + c \end{aligned}$$

Now the problem is to solve these three simultaneous equations. Techniques from elementary algebra or linear algebra (matrices) readily give the result. Unfortunately, if we have many points, there are many simultaneous equations and the calculations become tedious.

Another method of finding a polynomial curve through given points is due to the mathematician Lagrange and has advantages from the computational point of view. Lagrange used the combination of points approach to find a parametric description for the curve. Using the three points given above, we want to find blending functions $\alpha_0(t)$, $\alpha_1(t)$, and $\alpha_2(t)$ so that

$$P(t) = \alpha_0(t)P_0 + \alpha_1(t)P_1 + \alpha_2(t)P_2$$

In order for the curve $P(t)$ to go through P_0 there must be some value of t that makes $P(t) = P_0$. Let's arbitrarily decide that the curve should go through P_0 when $t = 0$. So $P(0) = P_0$. Similarly, we can decide that when $t = 1$ the curve passes through P_1, and when $t = 2$ the curve passes through P_2, so $P(1) = P_1$ and $P(2) = P_2$.

Now we need to pick the blending functions $\alpha_i(t)$, $i = 1, 2, 3$. Since $P(0) = P_0$, we would like $\alpha_0(0) = 1$, $\alpha_1(0) = 0$, and $\alpha_2(0) = 0$. After taking the similar conditions for P_1 and P_2 into account, it is clear that we would like $\alpha_0(t)$ to satisfy

$$\alpha_0(0) = 1,\ \alpha_0(1) = 0,\ \alpha_0(2) = 0.$$

If we assume that $\alpha_0(t)$ is zero only when $t = 1$ or $t = 2$, then we know that $\alpha_0(t) = K(t-1)(t-2)$, for some constant K. Since $\alpha_0(0) = 1$, we must have $K = 1/2$. So

$$\alpha_0(t) = \frac{(t-1)(t-2)}{2}.$$

Using the same procedure, we find

$$\alpha_1(t) = -t(t-2),$$
$$\alpha_2(t) = \frac{t(t-1)}{2}.$$

Now we have the parametric description of the curve,

$$P(t) = \frac{(t-1)(t-2)}{2}P_0 - t(t-2)P_1 + \frac{t(t-1)}{2}P_2.$$

If we substitute the particular points given above, we get

$$x(t) = -t^2 + 4t - 1$$
$$y(t) = 4t^2 - 10t + 9.$$

This curve is a parabola (see Exercise 6), but it is a different parabola than the one of the form $y = ax^2 + bx + c$. The axis of the Lagrange parabola is not parallel to the y-axis. The two parabolas are sketched in Figure 2. If we just want the curve segment that starts at P_0, passes through P_1, and ends at P_2, then we can restrict t to the interval $[0, 2]$.

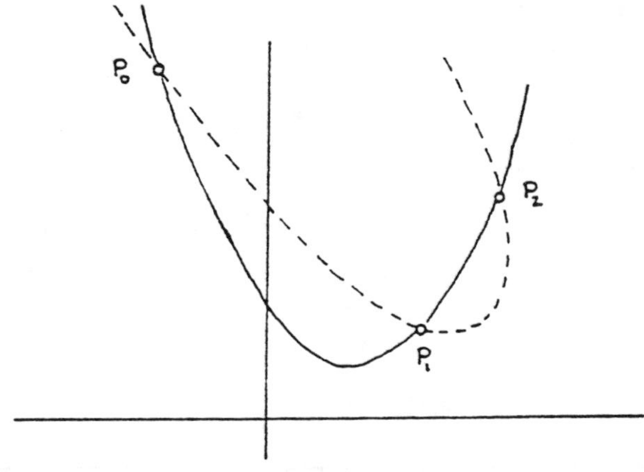

Figure 2.

Using Lagrange interpolation, we can quickly write down the parametric description of a polynomial curve which passes through any number of given points. The method is usually easy and fast for a computer to execute. To review, the method singles out integer values for t $(0, 1, 2, \cdots, n)$, and then picks blending functions so that $\alpha_i(i) = 1$ and $\alpha_i(t) = 0$ for the other selected integer values. This means that for $i = 0, \cdots, n$

$$\alpha_i(t) = \prod_{\substack{k=0 \\ k \neq i}}^{n} \frac{(t-k)}{(i-k)}$$

where the symbol $\prod$ means product. The product is taken over integers k from $k = 0$ up to $k = n$, skipping over $k = i$. For instance, in the example we just did,

$$\alpha_1(t) = \frac{(t-0)(t-2)}{(1-0)(1-2)} = -t(t-2).$$

It might seem that interpolation is all we need for our design system. However, there is an unfortunate problem. The more points we specify, the higher the degree of the polynomials in the parametric description, and hence the more "wiggles" there may be in the curve. For example, if we specify ten points, then the resulting parametric description will involve ninth degree polynomials. Calculus helps determine just how wiggly these polynomials can be. We find the local maxima and minima of a function by finding zeroes of the first derivative. For a ninth degree polynomial, the first derivative is a polynomial of degree eight. From the Fundamental Theorem of Algebra we know that a polynomial of degree eight can have as many as eight real zeroes. This implies that if the parametric description involves ninth degree polynomials, there can be as many as eight maxima and minima for both the x and y coordinates—a rather wiggly curve!

To see this wiggle problem in action, look at Figure 3. In Figure 3a, there are five control points and a dotted line indicating the desired design curve through them. Figure 3b shows the actual fourth degree polynomial that interpolates the five points. It is too wiggly to be useful for design here.

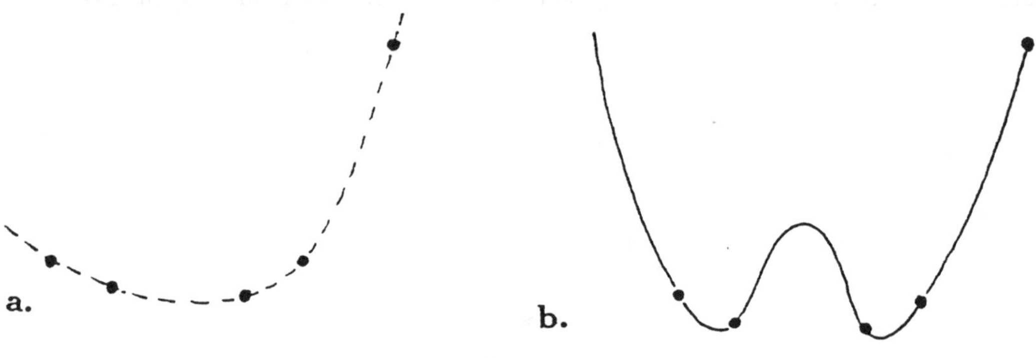

Figure 3.

Exercises

5. Solve for a, b, c to find the explicit description of the quadratic polynomial whose graph goes through the points $(-1, 9)$, $(2, 3)$ and $(3, 5)$.

6. Suppose a curve is given implicitly as

$$Ax^2 + Bxy + Cy^2 + Dx + Ey + F = 0$$

It is shown in analytic geometry that if $B^2 - 4AC = 0$ then this curve is a parabola. Starting from the parametric equations we derived for the Lagrange curve through $(-1, 9)$, $(2, 3)$ and $(3, 5)$, simplify by combining the equations to eliminate t^2 and then solve for t in terms of x and y. Substitute this into the expression for $y(t)$ to get an implicit form like the one above. Then show that the curve is indeed a parabola.

7. Use the Lagrange interpolation method to find the polynomial curve through $(0, 4)$, $(1, 2)$, $(3, -1)$, $(5, 5)$.

Smoothness

One way around the "wiggle" problem is to avoid interpolating many points and instead only pick a few at a time. For example, if we have points P_0 through P_6, then we need a sixth degree polynomial to interpolate all the points. Suppose now that we take only the first four points P_0 through P_3, and interpolate them with a segment of a cubic polynomial. Then take P_3 through P_6, and interpolate again with a cubic segment. Since the two segments both go through P_3, we can simply patch the two segments together to interpolate all the points. It sounds easy, but there is a catch. The resulting patched-together curve may not be very smooth at the point P_3. Figure 4 shows what can happen. The slope of the first segment at P_3 is different than the slope of the second segment there. For smoothness, the derivatives should match at the junction point.

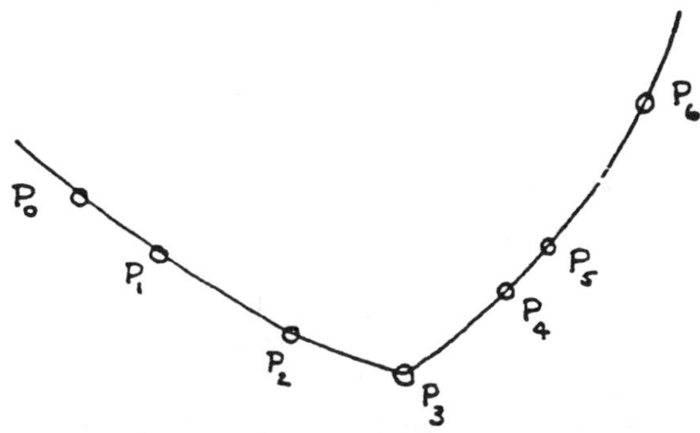

Figure 4.

Incidentally, we can patch together curves in a variety of ways. Taking two points at a time gives a curve that is just a sequence of line segments (called *piecewise linear*), and using three points at a time means we put together parabolic segments. In turns out from experience that cubic curve segments tend to be flexible enough for artistic concerns and yet of low enough degree to avoid too much wiggle. Because of this, cubic segments are most often used in computer design systems.

Calculus is essential in understanding what it means for the patched-together curve segments to form a smooth curve. First, since the end point of one segment is the initial point of the next, the curve is at least continuous. We call this *zero-order smoothness.* Now if there is a corner at the junction of two segments, then the slope of the curve has jumped at the junction point. This means that the derivative of our patched-together curve is not continuous at the junction point. To avoid corners, we need to insure that the derivative is continuous. This is called *first-order smoothness.*

It is possible that the first derivative is continuous, but the second derivative is not continuous. In this case the curve will look fairly smooth, but there will be sections where it bends a little too awkwardly. To avoid this, we could try to guarantee that the second derivative is continuous at all junction points in our patched-together curve. This is called *second-order smoothness.* Second-order smoothness is very important in designing ship hulls where abrupt bends can cause excessive water drag on the ship, and in designing parts of a car where air drag is a factor. In most design tasks, however, first-order smoothness is the prime concern, so we will concentrate on it.

Our first task is to make certain we can calculate the derivative. Since the computer screen effectively works with the x-y coordinate system, the derivative we need is usually dy/dx (sometimes we may prefer dx/dy). From the chain rule, we have

$$\frac{dy}{dt} = \frac{dy}{dx} \cdot \frac{dx}{dt}.$$

Hence we find that if $dx/dt \neq 0$, then

$$\frac{dy}{dx} = \frac{dy/dt}{dx/dt}.$$

For example, suppose we have the parametric curve

$$x(t) = t^2 - 3t, \quad y(t) = t^3 + t.$$

Then

$$\frac{dy}{dx} = \frac{3t^2 + 1}{2t - 3}.$$

When $t = 0$, we have $x = 0$ and $y = 0$, so the point (0,0) is on the curve. At this point, the slope is $-1/3$. Notice that $dx/dt = 0$ at the point $t = 3/2$. At this point the tangent line is vertical.

Exercises

8. Consider the parametric description we developed for a line segment: $P(t) = (1-t)P_0 + tP_1$. Find the derivative using this description and show that it is indeed the slope of the line between the two points.

9. The curve segment given in Exercise 2 forms a loop as t goes from 0 to 1. This means that the tangent line must be vertical at some point. Find this point.

10. Suppose we build a curve by taking pieces of the graphs of two functions. For $x \leq 0$, let $y = x^3 + 2x$, and for $x \geq 0$, let $y = x^2 + 2x$. Show that the resulting curve has zero-order smoothness and first-order smoothness, but not second-order smoothness.

Bézier Curves

In the early 1960's, two engineers working in the French automotive industry, P. Bézier and P. de Casteljau, independently developed a method for designing curves that solved most of the wiggle and smoothness problems with which engineers had wrestled. Bézier worked for Renault and de Casteljau worked for Citroën, so at first their results were considered manufacturing secrets. The work of de Casteljau was slightly earlier than Bézier's, but since it was never widely published—and Bézier's work was—the curves produced with this technique are now called *Bézier curves.*

The method developed by these two designers relies on taking a combination of the control points. Unlike the Lagrange interpolation method, where the blending functions are chosen so the curve goes through all the control points, the Bézier method chooses blending functions in a different way. To understand the motivation for the particular choices, it helps to look first at a geometric construction due to de Casteljau.

Figure 5a shows three initial control points labeled P_0, P_1, P_2. Experience with drafting led to the following construction of a curve. For any $0 \leq t \leq 1$, plot the points

$$B_0(t) = (1-t)P_0 + tP_1,$$
$$B_1(t) = (1-t)P_1 + tP_2.$$

Notice that as t goes from 0 to 1, $B_0(t)$ goes from P_0 to P_1, and $B_1(t)$ goes from P_1 to P_2. Now construct the point $B(t)$ on the line between $B_0(t)$ and $B_1(t)$ by

$$B(t) = (1-t)B_0(t) + tB_1(t).$$

As t goes from 0 to 1, $B(t)$ goes from $B_0(t)$ to $B_1(t)$. The Bézier curve $B(t)$ is shown in Figure 5b. It is not hard to imagine a mechanical linkage which would produce this construction.

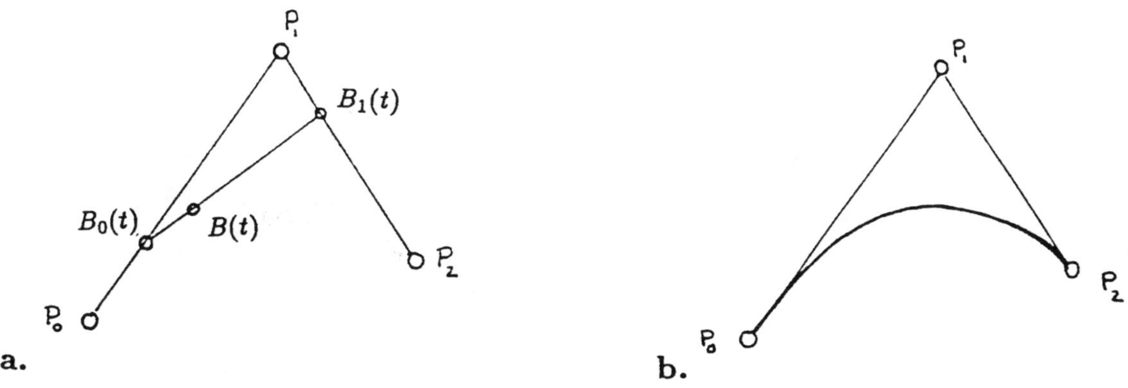

Figure 5.

The geometric construction is fine for a draftsperson, but is awkward for use on the computer. However, to derive the parametric description, all we need to do is substitute the expressions for $B_0(t)$ and $B_1(t)$ into the expression for $B(t)$. The result is

$$\begin{aligned} B(t) &= (1-t)[(1-t)P_0 + tP_1] + t[(1-t)P_1 + tP_2] \\ &= (1-t)^2 P_0 + 2t(1-t)P_1 + t^2 P_2 \end{aligned}$$

The Bézier curve is a combination of the three control points, and in fact $B(0) = P_0$ and $B(1) = P_2$ so the curve interpolates the first and last control point. Notice that the coordinate functions $x(t)$ and $y(t)$ for the curve are both quadratic functions of t. The curve is actually a parabola.

Consider the same type of geometric construction done this time with four initial control points. Figure 6 shows the construction. Again, for t in the interval [0,1], we proceed to find intermediate points and line segments. This time we must go one more step before we find the single point $B(t)$. First we place $B_0^1(t)$, $B_1^1(t)$, and $B_2^1(t)$ on the line segments between the control points. Then we place $B_0^2(t)$ and $B_1^2(t)$ on the line segments between the points first placed. We have

$$\begin{aligned}B_0^1(t) &= (1-t)P_0 + tP_1\\ B_1^1(t) &= (1-t)P_1 + tP_2\\ B_2^1(t) &= (1-t)P_2 + tP_3\\ B_0^2(t) &= (1-t)B_0^1(t) + tB_1^1(t)\\ B_1^2(t) &= (1-t)B_1^1(t) + tB_2^1(t),\end{aligned}$$

and finally,

$$B(t) = (1-t)B_0^2(t) + tB_1^2(t).$$

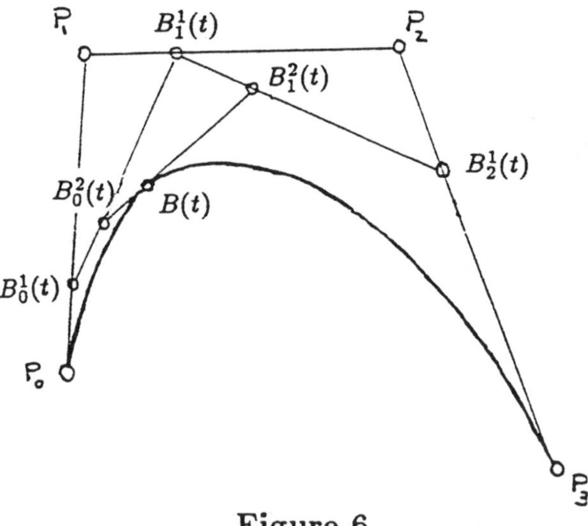

Figure 6.

Again we can simplify the expression, but it is a little easier if we recognize the recurrence relation

$$B_i^r(t) = (1-t)B_i^{r-1}(t) + tB_{i+1}^{r-1}(t).$$

Now notice that $B_0^2(t)$ is really the Bézier curve with control points P_0, P_1, P_2, and $B_1^2(t)$ is the Bézier curve with control points P_1, P_2, P_3. Since we previously derived the expression for three control points, we can use it now to get the expression for four points.

$$\begin{aligned}B(t) &= (1-t)B_0^2(t) + tB_1^2(t)\\ &= (1-t)[(1-t)^2P_0 + 2t(1-t)P_1 + t^2P_2] + t[(1-t)^2P_1 + 2t(1-t)P_2 + t^2P_3]\\ &= (1-t)^3P_0 + 3t(1-t)^2P_1 + 3t^2(1-t)P_2 + t^3P_3.\end{aligned}$$

There is no need to stop at four points. If we have any number of points, we can use the recursive technique to write down the Bézier curve as a combination of all the points. You may have already seen the pattern in the blending functions. The general formula for the Bezier curve given by the $n+1$ control points $P_0, \ldots, P_n$ is

$$B(t) = \sum_{i=0}^{n} \binom{n}{i} t^i(1-t)^{n-i}P_i \qquad 0 \le t \le 1,$$

where $\binom{n}{i} = \dfrac{n!}{i!(n-i)!}$ is a binomial coefficient. The blending functions $\binom{n}{i}t^i(1-t)^{n-i}$ which give us Bézier curves occurred earlier in mathematical literature, and are called *Bernstein polynomials*.

Figure 7 shows several Bézier curves and their control points.

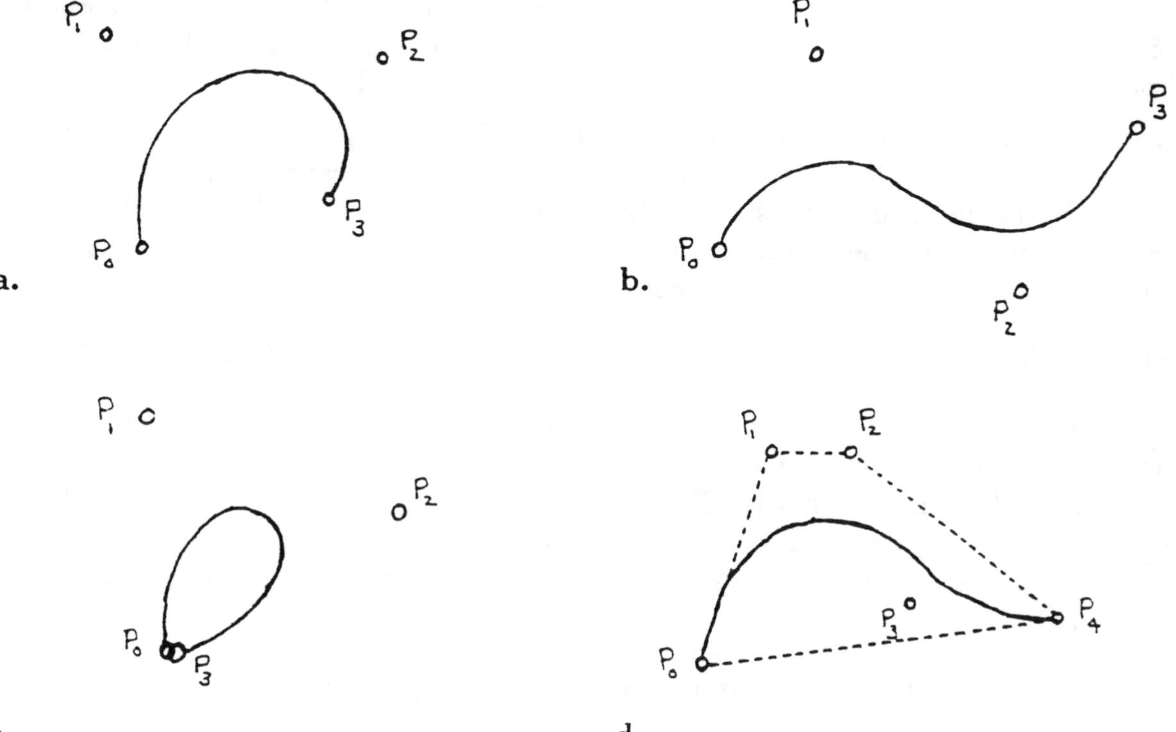

Figure 7.

Exercises

11. Using the recurrence relation, derive the blending functions for a Bézier curve with five control points (i.e. Bernstein polynomials of degree 4). Verify that the formula given for $B(t)$ is correct in this case.

12. Consider the ith Bernstein polynomial of degree n, *displaystyle*$\binom{n}{i}t^i(1-t)^{n-i}$. Show that the unique maximum for this polynomial in the interval $0 \leq t \leq 1$ occurs at $t = i/n$.

13. If you have access to computer software for drawing Bezier curves, experiment with drawing Bezier curves with five or six control points. In particular, study how the curves change as some of the control points are moved. Turn in your most interesting experiment.

Properties of Bézier Curves

Bézier curves have a number of very nice properties. From the definition of $B(t)$, it is clear that $B(0) = P_0$ and $B(1) = P_n$.

Property 1. *The Bézier curve interpolates the first and last control points.*

This property implies that if we set $P_0 = P_1$, then the resulting curve is a closed curve. See Figure 7c.

The polygon indicated by the dotted line in Figure 7d encloses all of the control points for the Bézier curve. This polygon is convex, meaning that there are no corners that point in. In fact the polygon is the smallest convex polygon which includes all the control points, and we refer to it as the *convex hull* of the control points.

The de Casteljau algorithm for producing the Bézier curve picks points on line segments between control points or previously constructed points. Because of this procedure, points on the Bézier curve are always inside the convex hull of the control points. This is a nice property from the designer's standpoint since it guarantees that the curve is contained in a region that can be specified with the control points.

Property 2. *The Bézier curve lies entirely within the convex hull of the control points.*

Bézier curves solve the wiggle problem to a certain degree. Since the curve can be constructed by picking points on line segments between existing points, it doesn't tend to wiggle very much. It turns out that the Bézier curve cannot intersect a given straight line any more times than the polygonal curve formed by connecting the control points in order. Figure 8 shows an example; the curve is less wiggly than the polygon. For a proof of this fact, see Farin (1988).

Property 3. *The Bézier curve is less wiggly than the polygon formed by the control points.*

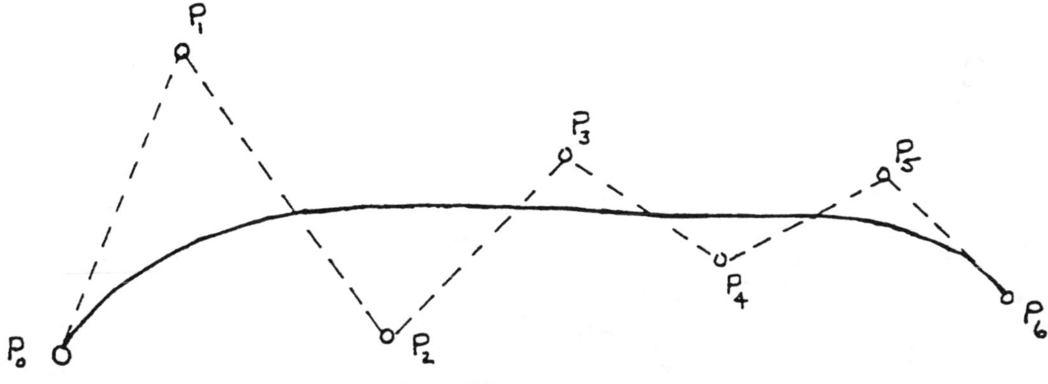

Figure 8.

In a complex design, we usually want to patch together several curve segments to get the final curve. Our design system should insure that we have first-order smoothness at the junction points. To see how the Bézier curve addresses this problem, first notice that taking the derivative and setting $t = 0$ gives

$$B'(0) = -nP_0 + nP_1 = n(P_1 - P_0)$$

(see Exercise 14). This means that if the coordinates of P_0 are (x_0, y_0) and the coordinates of P_1 are (x_1, y_1) then

$$x'(0) = n(x_1 - x_0), \quad y'(0) = n(y_1 - y_0).$$

Hence the value of the derivative at $t = 0$ is

$$\frac{y'(0)}{x'(0)} = \frac{y_1 - y_0}{x_1 - x_0}.$$

This is just the slope of the line segment from P_0 to P_1. When $t = 0$, the curve is at the point P_0, so the slope of the curve at the first control point is the same as the slope of the line segment P_0P_1. Similarly, the slope of the curve at the last control point P_n is the same as the slope of the line segment $P_{n-1}P_n$.

Property 4. *The slope of the Bézier curve at P_0 is the same as the slope of the line segment P_0P_1, and the slope of the curve at P_n is the same as the slope of the line segment $P_{n-1}P_n$.*

This property of the slopes is very convenient for the designer. For example, suppose we have two curve segments with control points P_0, P_1, P_2, P_3, and Q_0, Q_1, Q_2, Q_3. To make certain the segments meet, we set $P_3 = Q_0$. To insure that the junction is first order smooth, we simply make sure that P_2, P_3 $(= Q_0)$, and Q_1 are all on the same line. This guarantees that the slopes match up. Figure 9 shows the situation.

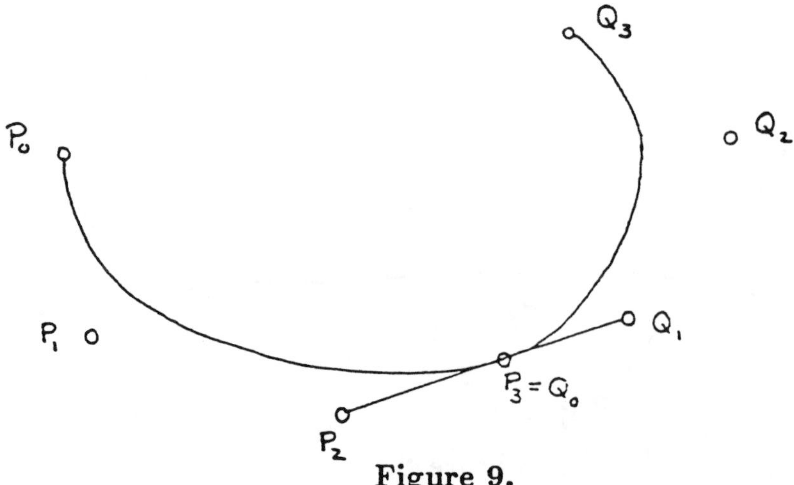

Figure 9.

There is one other property of Bézier curves that guarantees they are fairly well-behaved. If all our control points lie on a straight line, we expect the Bézier curve to be on the same straight line. This is indeed the case. In fact, it follows from Property 2.

Property 5. *If all the control points lie on a straight line, then the Bézier curve is on the same straight line.*

Exercises

14. Write out the first few terms of $B(t)$. Differentiate, and verify that $B'(0) = -nP_0 + nP_1$.

15. Show that the Bézier curve with control points (1,2), (2,4), (5,10) is on the straight line through these points.

The Final Design System

It should be clear now that if we have a computer design system that allows us to do Lagrange interpolation and Bézier curves, we can certainly design car bodies. The resulting curves have a compact mathematical description so there should be little problem in communicating the design to the manufacturing engineers.

Most computer-aided design (CAD) systems include some curve sketching facility, usually Bézier curves. With the more sophisticated systems, the user can place control points, look at the resulting curve, and then alter individual control points until the curve looks right. This interactive approach to design is invaluable. In addition to widespread use in industrial design, CAD systems are valuable in many other areas. For example, commercial animators can use CAD to help make films if they can describe the motion curves for their characters, and meteorologists use CAD to draw isobars on weather maps.

References

Angel, Edward (1990), *Computer Graphics*, Addison-Wesley.

Farin, Gerald (1988), *Curves and Surfaces for Computer Aided Design*, Academic Press.

Hill, Francis, Jr. (1990), *Computer Graphics*, Macmillan.

Pokorny, C. and Gerald, C. (1989), *Computer Graphics: The Principles Behind the Art and Science*, Franklin, Beedle, & Associates.

Answers to Exercises

1. $\frac{x^2}{4}+\frac{y^2}{9}=\frac{4\cos^2 t}{4}+\frac{9\sin^2 t}{9}=1.$
2.

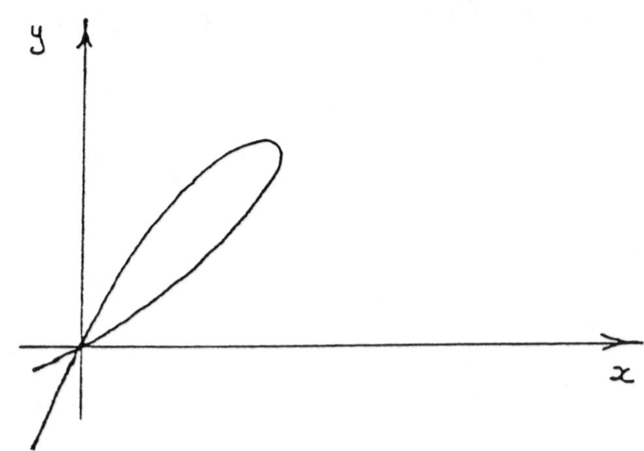

3. In both cases, $y=x^2$ for $0\le x\le 1$. However, in the first case, when $t=1/2$, we get the point $(1/4,1/16)$, whereas in the second case we are at the point $(1/2,1/4)$. For the larger interval, the first segment is only half of the second.
4. We have $\alpha_1(t)=1-\alpha_0(t)$. Therefore, $P(t)=\alpha_0(t)P_0+(1-\alpha_0(t))P_1$. This gives $x(t)=\alpha_0(t)x_0+(1-\alpha_0(t))x_1$ and $y(t)=\alpha_0(t)y_0+(1-\alpha_0(t))y_1$. Now solving for $\alpha_0(t)$ in the expression for $x(t)$ and substituting into the expression for $y(t)$ gives

$$\begin{aligned} y &= \frac{x-x_1}{x_0-x_1}\cdot y_0+(1-\frac{x-x_1}{x_0-x_1})y_1 \\ &= (\frac{y_0-y_1}{x_0-x_1})\cdot x+(y_1-x_1\cdot\frac{y_0-y_1}{x_0-x_1}) \\ &= mx+b \end{aligned}$$

5. $a=1$, $b=-3$, $c=5$
6. $t=\frac{1}{6}(y+4x-5)$. This gives $A=1/36$, $B=8/36$, $C=16/36$.
7. $P(t)=\frac{(t-1)(t-2)(t-3)}{-6}P_0+\frac{t(t-2)(t-3)}{2}P_1+\frac{t(t-1)(t-3)}{-2}P_2+\frac{t(t-1)(t-2)}{6}P_3.$
   This gives

$$\begin{aligned} x(t) &= -\frac{1}{6}t^3+t^2+\frac{1}{6}t \\ y(t) &= \frac{5}{3}t^3-\frac{11}{2}t^2+\frac{11}{6}t+4. \end{aligned}$$

8. $x'(t) = -x_0 + x_1$ and $y'(t) = -y_0 + y_1$. Hence $dy/dx = \dfrac{y_1 - y_0}{x_1 - x_0}$. This is the slope of the line between P_0 and P_1.

9. $x'(t) = 1 - 2t$. Therefore the tangent is vertical when $t = 1/2$. This occurs at the point $(1/4, 3/8)$.

10. The problem point is at the junction: (0,0). The first segment's slope here is 2 and second derivative is equal to 0. The second segment's slope is also 2, but the second derivative is 2.

12. If i is not 0 or n, the derivative is $\binom{n}{i}[it^{i-1}(1-t)^{n-i} - t^i(n-i)(1-t)^{n-i-1}]$. Setting this equal to zero gives $t = 0$ or $t = 1$ or $t = i/n$. Note that $t = 0$ and $t = 1$ give values of zero, but $t = i/n$ gives a positive value; hence it is the maximum. The cases $i = 0$ and $i = n$ must be checked separately.

14. $B(t) = (1-t)^n P_0 + nt(1-t)^{n-1}P_1 + \dfrac{n(n-1)}{2}t^2(1-t)^{n-2}P_2 + \ldots$ When we differentiate, the only terms which don't contain t are $-n(1-t)^{n-1}P_0 + n(1-t)^{n-1}P_1$.

15. $x(t) = 2t^2 + 2t + 1$ and $y(t) = 4t^2 + 4t + 1$. Hence $y = 2x$.

MODELING THE AIDS EPIDEMIC

Author: Steven Janke, Colorado College, Colorado Springs, CO 80903

Calculus Needed: Separable differential equations, exponential and logarithm functions, integral as a limit of Riemann sums, integration by partial fractions, improper integrals.

Areas of Application: Biology, medicine.

The Problem: Understanding the Growth of AIDS

Acquired Immune Deficiency Syndrome (AIDS) has cast a pall on all our lives. The first cases were observed in the early 1980's, and by the end of 1991 about 200,000 people in the United States had been diagnosed with this terrible disease.† It is frightening because it is deadly, and we have no cure. While medical researchers struggle to stop the spread of the virus, the Centers for Disease Control (CDC) carefully record the number of diagnosed cases. With these data, researchers can look at the progress of the disease and search for important patterns. Is the epidemic growing like epidemics in the past? Will it eventually die out on its own? How many cases will be diagnosed next year, so that we can plan for adequate medical care? How can we tell whether the massive information campaign about AIDS is having an affect on the spread of the disease?

Epidemics of the Past

The first major epidemic for which we have records was the "Plague of Justinian." In 541 A.D. the Byzantine empire was attacked by what was probably a mixture of plague, smallpox, and measles. The epidemic lasted for more than fifty years. The notorious "Black Death" (bubonic plague) spread through Central Asia and Europe starting in 1338. This epidemic killed more people in four years than the Plague of Justinian killed in fifty years. Plague devastated London in 1665, spread through England and closed Cambridge University, sending Isaac Newton home to discover the law of gravity.

In this century, there have been several epidemics. One of the worst was the 1918 influenza epidemic, which spread nearly everywhere and killed 22 million people. Poliomyelitis held the attention of medical researchers throughout the first half of this century until

† There is also a serious worldwide epidemic of AIDS. However, because of behavioral and health care differences, the pattern of the epidemic is quite different in Africa, say, from in the United States. In this module we will focus on modeling the AIDS epidemic in the U.S. A good introduction to models of the global epidemic is Anderson and May (1992).

a vaccine in 1955 finally stopped the epidemic. In the last decade we have finally seen smallpox, which was responsible for many epidemics in the past, effectively eradicated.

In 1760, when smallpox was epidemic in Europe, Daniel Bernoulli produced the first mathematical model of epidemics. Bernoulli was interested in modeling smallpox in order to assess whether the new technique of innoculating people with weaker or dead viruses would halt its spread. In some of his work, he investigated the logistic growth curve, which we will see in the next section.

AIDS is the current horror, but is not the first sexually transmitted disease to become epidemic. Syphilis and gonorrhea among others have been epidemic in the past and remain endemic (i.e. present, but in control) today. As we will see, there has been a surprise with AIDS. Unlike classic epidemics, the number of cases has not been growing exponentially.

The Logistic Model

As a first step toward understanding the growth of the AIDS epidemic, let us consider the simplest mathematical model of the growth of an epidemic. We are interested in the number of infected people at time t, call it $R(t)$. In this first model, we assume that the growth rate of the number of infected people is proportional to the current number of infected people:

$$\frac{dR(t)}{dt} = kR(t). \tag{1}$$

Here k is a constant of proportionality and represents how readily the disease can be spread. This differential equation can be easily solved to get

$$R(t) = R_0 e^{kt} \tag{2}$$

where R_0 is the number of infected people at time zero. Remember that there are infinitely many solutions to the differential equation (1) above. It requires an initial condition (in this case the initial number of infecteds) to determine the unique solution.

Our model has led to a simple function giving the number of infected people at time t. We can apply the model to predict how many people will have the disease at various times. It turns out that if we consider, for example, the spread of influenza in a small town, our function gives fairly accurate values for small values of t. However, as time goes on, the model makes steadily worse predictions. The reason is clear: as more people become infected, it is harder to find people to infect.

Hence our first model must be altered. This time we argue that the rate of growth depends not only on how many are currently infected, but also on how many are uninfected (susceptible). If $R(t)$ is again the total number of infected people at time t and if there is a total of N people in the population, then the number of susceptible people is $N - R(t)$. The number of possible contacts between infecteds and susceptibles is $R(t)(N - R(t))$, so we now assume that the epidemic's rate of growth is proportional to this product. The differential equation becomes

$$\frac{dR(t)}{dt} = kR(t)(N - R(t)). \tag{3}$$

The constant of proportionality k depends on the likelihood of contact and the probability that contact leads to transmission of the disease.

This differential equation is a little more difficult to solve. Start by separating the variables:

$$\frac{dR(t)}{R(t)(N - R(t))} = k\ dt. \tag{4a}$$

Now integrate both sides. The right-hand side is easy to integrate, but the left-hand side requires using partial fractions (See Exercise 2). The result is

$$\ln \frac{R(t)}{N - R(t)} = Nkt + C, \tag{4b}$$

where C is a constant. Applying the exponential function to both sides gives

$$\frac{R(t)}{N - R(t)} = \overline{C}e^{Nkt}. \tag{5}$$

In this equation, $\overline{C} = e^C$ is currently an unknown constant. However, if we know how many people are infected at time zero, say R_0, then by setting $t = 0$, we can solve for $\overline{C}$. Then a little more algebra allows us to solve for $R(t)$. The result is

$$R(t) = \frac{NR_0}{R_0 + (N - R_0)e^{-kNt}}. \tag{6}$$

This type of growth is called *logistic growth.* The term kN in the exponent is the parameter that determines how quickly this curve initially rises. We will call this term the *growth factor* for the curve. There are two important facts about the behavior of $R(t)$ that we need to note.

Small t: If R_0 and t are small, then $(N - R_0)$ is close to N and e^{-kNt} is not close to zero. This makes $(N - R_0)e^{-kNt}$ much bigger than R_0, so we have

$$R(t) \approx \frac{NR_0}{Ne^{-kNt}} = R_0e^{kNt}$$

For small t, logistic growth looks like exponential growth.

Large t: $\lim_{t\to\infty} R(t) = N$, so that $R(t)$ does not grow indefinitely, but rather approaches the size of the population. This makes sense since the most infecteds we can ever have is N.

The graph of $R(t)$ is shown in Figure 1. The graph of the derivative $R'(t)$ (see Exercise 4) is often called the *epidemic curve.* This curve tells us how fast the epidemic is growing at any particular time. For logistic growth, the epidemic curve increases to a maximum value and then decreases to zero.

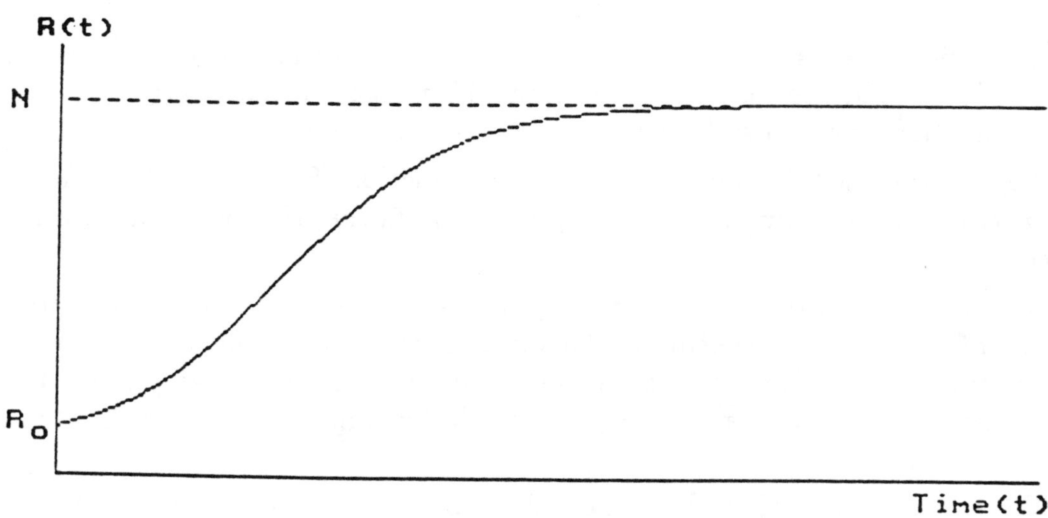

Figure 1. The Logistic Curve.

Exercises

1. Assume that we have exponential growth as in equation (2). Suppose that 50 people are initially infected, and that after one week, 70 people are infected. How long is it before 100 people are infected? How much longer will it be until 200 people are infected? The time interval you are seeking is called the *doubling time.* Show that no matter where we start on the growth curve, the doubling time is constant.
2. Integrate both sides of equation (4a) to get equation (4b). (The lefthand integral requires a partial fraction decomposition.)
3. Derive equation (6) from equation (5).
4. Find the equation for the epidemic curve and determine when it reaches its maximum. Use a computer to graph the epidemic curve and the logistic curve on the same axes.
5. Consider a small town of 5000 residents. Ten residents get a cold after a trip to another town. Compare the exponential model for viral infection growth with the logistic model. Assume the growth factor for both models is 1.2 (i.e. k in the exponential model and kN in the logistic model). Use both models to calculate how many people will have a cold at time 1 week and at time 4 weeks. (We are assuming that no one recovers.)
6. What happens if the growth factor for the logistic model is negative?

Assumptions in the Logistic Model

The logistic model is a simple model for the growth of an epidemic—probably too simple to apply to AIDS. Let us list the main simplifying assumptions in the logistic growth model, and mention some possible criticisms of those assumptions.

1. We have assumed that $R(t)$ is a continuous function. Since we are counting people, this is not strictly true. However, if the population is fairly large, assuming continuity is reasonable.

2. We have assumed that the growth rate is proportional to the product of the numbers of infecteds and susceptibles. This makes sense if we think of the product as representing the number of possible contacts between infecteds and susceptibles. However, we must be aware that there may be some other relationship between growth rate and the number of infecteds.

3. We have assumed that there are only two categories for people: infecteds and susceptibles. It is possible that some are immune, that some die, or that once people recover from the disease they become immune.

4. We have assumed that once a contact results in an infection, the newly infected person immediately develops the disease. However, there may be a *latency period* between the time of infection and the onset of the disease. This will be especially important for AIDS.

5. We have assumed that any person can infect any other person. This implies that all people behave in the same way and there is homogeneous mixing in the population. This is a critical assumption and we will refer to it as the *homogeneous assumption.* Clearly in real populations there are vastly different behaviors. The differences are particularly important in modeling sexually transmitted diseases like AIDS.

A Portrait of AIDS

In order to be more specific in our modeling effort, we need to understand a little about the development of AIDS. AIDS is caused by the HIV virus which is usually transmitted in adults by one of three methods: blood transfusions, needle-sharing by drug users, and sexual contact. Since careful testing has increased the safety of the blood supply, transmission by blood transfusions has decreased considerably. The model in the next section will concentrate on transmission by sexual contact.

Once a person has been infected, it takes from two to six months for antibodies to appear in the bloodstream. At this point, the patient has an HIV infection. During this time, the amount of free virus in the bloodstream peaks dramatically. The amount of free virus then subsides as the virus hides in other cells and carries on its relentless destruction of immune system cells. The patient moves into a latent period that can last as little as 2 years, or up to 15 years or perhaps even longer. Then the free virus begins increasing again, and because of natural selection it tends to be much more virulent. At some stage the patient begins to suffer from fever and other symptoms. This stage is referred to as AIDS Related Complex (ARC). Finally, as the immune system begins to fail, opportunistic infections take over. The patient usually contracts pneumonia, Kaposi's sarcoma (a cancer), or a variety of other diseases that gradually take away life. At this final stage the patient has full-blown AIDS.

For our modeling effort, first note that we would expect sexual behavior to be a critical factor. Since behavior varies widely, the homogeneous assumption—number 5 above—is probably not warranted. Secondly, note that there is a considerable latency period with HIV infection, so that assumption 4 does not hold. Furthermore, not everyone develops AIDS at the same time after infection. This is important, since it is not clear that we know how many infective people there are at a given time. In fact we do not have good estimates of how many people are infected with HIV. The data we have chronicles full blown AIDS.

The CDC keeps track of the diagnosed cases of AIDS. These are people that have entered the last stage of the disease. Table 1 gives the data as of the end of August 1991. As you might suspect, data are not reported promptly to the Centers, so there are constant updates and the table will undoubtedly change slightly over time.

Using statistical techniques, we can determine which kind of function best fits the data. It turns out that a cubic function fits better than an exponential function. Specifically, $R(t) = 187(t - 1981)^3 - 274$ fits the data well. Since we are near the beginning of the epidemic, we would expect $R(t)$ to be approximately exponential if the logistic model were a good model for AIDS growth. Since a cubic function fits better, we will need to develop a better model.

Note that the data are subject to all the pitfalls of real life data collection. There are delays and some cases are not reported. Even so, the growth of AIDS is sufficiently different from exponential growth to cause us to improve our model.

Exercise

7. If the growth of AIDS were exponential, then we saw in Exercise 1 that the doubling time would be constant. Look at the data in Table 1 and see what is happening to the doubling time.

Table 1. The AIDS Epidemic.

The following data were reported in the August 1991 issue of *HIV/AIDS Surveillance* published by the Centers for Disease Control. The Cases column gives the number of AIDS cases diagnosed in the designated interval. (Remember that these are AIDS cases, not just HIV infections.)

Half-Year		Cases	Cumulative
1981	Jan-June	92	92
	July-Dec	203	295
1982	Jan-June	390	685
	July-Dec	689	1,374
1983	Jan-June	1,277	2,651
	July-Dec	1,642	4,293
1984	Jan-June	2,550	6,843
	July-Dec	3,368	10,211
1985	Jan-June	4,842	15,053
	July-Dec	6,225	21,278
1986	Jan-June	8,215	29,493
	July-Dec	9,860	39,353
1987	Jan-June	12,764	52,117
	July-Dec	14,173	66,290
1988	Jan-June	16,113	82,403
	July-Dec	16,507	98,910
1989	Jan-June	18,452	117,362
	July-Dec	18,252	135,614
1990	Jan-June	18,601	154,215
	July-Dec	16,636	170,851
1991	Jan-June	12,620	183,471

Notes:

1. This table includes only adults and adolescents. There have been 3199 cases reported among children less than 13 years old.
2. 85 cases were reported before 1981.
3. The table gives the number of AIDS cases diagnosed, not the number of deaths. The August issue reports that 118,411 individuals (adults, adolescents, and children) have died from AIDS.
4. The last two numbers in the Cases column are almost certainly too low, due to delayed reporting. They are omitted in the curve-fitting.
5. Both an exponential curve and a cubic curve give decent fits to the data, but the cubic curve explains 99.8 percent of the variance whereas the exponential curve explains 92.8 percent. It can also be argued from the pattern of residuals that the cubic fits better.

The Saturation Wave Model

The most important properties which prevent the logistic model from describing the growth of AIDS are the variable latency period for AIDS and the lack of homogeneous mixing in the population. Models which include both non-homogeneous mixing and an uncertain latency period usually lead to complicated systems of differential equations which cannot be solved exactly, so that solutions must be approximated numerically. However, a group of researchers at Los Alamos [Colgate *et al.*, 1989] was able to avoid most of this complexity by making suitable simplifying assumptions, while still including the latency period and heterogeneous behavior in their AIDS model. The resulting differential equation is not only solvable, but also predicts cubic growth. We will sketch the development of this remarkable model in six steps. It is a good illustration of the interplay between complexity and simplifying assumptions in modeling.

Let

$$H(t) = \text{ the cumulative number of HIV infections up to time } t,$$
$$A(t) = \text{ the cumulative number of AIDS cases up to time } t.$$

STEP 1: Latency period. We will describe the variable latency period for AIDS by a *probability density function* $L(t)$. This is a function such that the probability that the latency period for an HIV infected individual is between time τ and time $\tau + \Delta\tau$ is $\int_\tau^{\tau+\Delta\tau} L(t)\,dt$. Figure 2 shows that this probability is approximately $L(\tau)\Delta\tau$ when $\Delta\tau$ is small.

Current evidence indicates that the AIDS latency period is always between 2 and 18 years. This means that $L(t) = 0$ for $t < 2$ or $t > 18$, and $\int_2^{18} L(t)\,dt = 1$. Later, we will assume a specific form for $L(t)$, subject to these constraints.

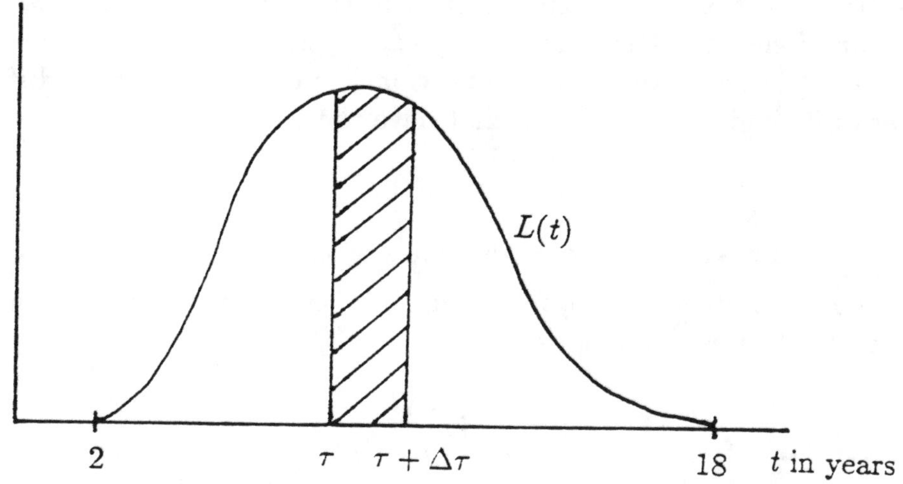

Figure 2. A Probability Density Function for the Latency Period.

STEP 2: Formula for $A'(t)$. Partition the time interval $[2, 18]$ into a large number of small subintervals $[\tau_i, \tau_i + \Delta\tau]$, $i = 1, \ldots, n$. Individuals will develop AIDS between time t and time $t + \Delta\tau$ if they were infected between time $t - \tau_i - \Delta\tau$ and time $t - \tau_i$, and their latency period was between τ_i and $\tau_i + \Delta\tau$. Hence we get the approximation

$$A(t + \Delta\tau) - A(t) \approx \sum_{i=1}^{n} [H(t - \tau_i) - H(t - \tau_i - \Delta\tau)] L(\tau_i)\, \Delta\tau. \tag{7}$$

Now divide both sides by $\Delta\tau$ and let $\Delta\tau \to 0$. The difference quotients approach derivatives, the Riemann sum approaches an integral, and we get

$$A'(t) = \int_2^t H'(t - \tau) L(\tau)\, d\tau. \tag{8}$$

STEP 3: Heterogeneous behavior. The Los Alamos group dealt with heterogeneous behavior by assuming the the population is divided into different risk classes, and people within each class tend to interact mostly with other people in that class.

They looked at studies of sexual behavior and found that if proportion of people is plotted versus number of sexual partners (p) per year, the function k/p^3 fits the data fairly well for $p \geq 1$. This means, as we might expect, that most people have about 1 partner per year and fewer people have 2 partners per year. Many fewer people have, say, 20 partners per year.

The Los Alamos group concluded that if there is a measure of risky sexual behavior, then the number of individuals with risk r, call it $N(r)$, should be inversely proportional to the cube of r. That is,

$$N(r) \approx \frac{N_0}{r^3} \tag{9}$$

where N_0 is some constant.

STEP 4: Growth in single risk group. Consider all individuals of risk r. We assume now that the HIV infection grows exponentially in this group with a growth rate that is proportional to r. Hence the HIV infection spreads faster among individuals who engage in riskier sexual behavior. If we let the proportionality constant be γ, and if we let $H_r(t)$ be the number of individuals with risk r that have the HIV infection, then

$$H_r(t) = H_r(0) e^{\gamma r t} \tag{10}$$

where $H_r(0)$ is the number of infected individuals when we start measuring time.

Since there are $N(r)$ individuals in the group, the entire group will be infected when $N(r) = H_r(0) e^{\gamma r t}$, which will happen at time

$$t = \frac{1}{\gamma r} \ln \frac{N(r)}{H_r(0)}. \tag{11}$$

We will call this time t the *saturation time.* The group is saturated if everyone is infected. The ratio $N(r)/H_r(0)$ is the number of people in a group divided by the number that are infected at the beginning of the epidemic. If we assume that this ratio is approximately constant over all r, we can conclude that the saturation time is inversely proportional to the risk r.

STEP 5: Saturation wave and HIV infection. Individuals who engage in riskier behavior are more likely to become infected early in the epidemic. One by one, the high risk groups become saturated before the lower risk groups: a *saturation wave* moves through the population from high risk groups to low risk groups. To approximate the number of individuals that have the HIV infection, we can add up all groups that are currently saturated. Suppose that the group with risk r_* just reached saturation. Then the number infected is approximately

$$\int_{r_*}^{\infty} N(r)dr = \int_{r_*}^{\infty} \frac{N_0}{r^3}\, dr = \frac{N_0}{2r_*^2}. \tag{12}$$

If we let t be the saturation time of group r_*, we know from Step 4 that $t = k/r_*$ where k is some constant. Substituting this in the righthand side of equation (12) gives Kt^2 where $K = N_0/2k^2$ is a constant. This is an approximation for $H(t)$, the number of individuals up to time t that have the HIV infection:

$$H(t) \approx Kt^2. \tag{13}$$

In other words, the HIV infection grows quadratically.

STEP 6: Cubic growth of AIDS. If the AIDS latency period lasts at least 2 years and not longer than 18 years, the simplest assumption is that the onset of AIDS is uniformly distributed in this interval. Hence we set

$$L(t) = \begin{cases} 0 & \text{if } t < 2, \\ \frac{1}{18-2} = 0.0625 & \text{if } 2 \le t \le 18, \\ 0 & \text{if } t > 18. \end{cases} \tag{14}$$

This means that the probability that an individual develops AIDS from 2 to 3 years after infection is 0.0625. The probability that AIDS begins from 2 to 4 years after infection is $2 \cdot 0.0625 = 0.125$.

Now return to equation (8) and use (13) and (14) to derive that, for $2 \le t \le 18$,

$$\begin{aligned} A'(t) &= \int_2^t H'(t-\tau) \cdot L(\tau)\, d\tau \\ &= \int_2^t 2K(t-\tau) \cdot 0.0625\, d\tau \\ &= 0.0625 \cdot K(t-2)^2. \end{aligned} \tag{15}$$

It is estimated that the HIV infection began in 1979, so if we are interested in the number of AIDS cases in the early 1990's, we are justified in taking $t \le 18$.

Now it is a simple matter to find $A(t)$:

$$\begin{aligned} A(t) &= \int_2^t A'(s)\,ds \\ &= \int_2^t 0.0625\,K(s-2)^2\,ds \\ &= 0.0625\,K \cdot \frac{1}{3}(t-2)^3. \end{aligned} \tag{16}$$

Equation (16) predicts that the cumulative number of AIDS cases is a cubic function of time.

You should note that this prediction comes at the end of a sequence of seemingly heroic assumptions (Exercise 8). It is quite sensitive to the shape of the risk curve $N(r)$ (Exercise 11), the uniform probability distribution for $L(t)$ (Exercise 10), and the restriction that $t \leq 18$ (Exercise 9). More importantly, it is grounded in a picture of HIV infection spreading exponentially through high risk groups of the population and then moving on to spread exponentially through lower risk groups. As such, the model is designed only for the early stages of the epidemic, and is not useful for predicting long-term effects. That's good, because it would otherwise predict that AIDS would become universal.

One particularly useful feature of the saturation wave model is its clear connection between $A(t)$ (AIDS cases) and $H(t)$ (HIV infections, which will become AIDS cases). Knowledge of $H(t)$ is crucial to estimating public health needs in the immediate future, but the data on $H(t)$ is so incomplete that our best hope for estimating it seems to be from models like this. You should certainly do Exercise 12, which asks you to use the data on $A(t)$ to estimate the 1992 value of $H(t)$. I think you will find the number sobering. On the other hand, you should remember the assumptions in the model from which it comes.

Exercises

8. List the major assumptions made in the saturation wave model.
9. Show that for $t > 18$, $A'(t) = 0.0625\,(H(t-2) - H(t-18))$, and that this leads to quadratic growth for $A(t)$.
10. Suppose that $L(t) = \frac{1}{128}(t-2)$ for $2 \leq \tau \leq 18$. This means that as the latency period becomes longer, the probability of developing AIDS becomes greater. Now calculate $A(t)$.
11. Determine what $N(r)$ would have to be in order for AIDS to grow exponentially.
12. Notice that K in equation (16) is the same K that is in equation (13). Also, recall that the curve $187(t-1981)^3 - 274$ fits the data in Table 1 and is hence an estimate for $A(t)$. Now estimate K by equating the coefficient of t^3 in the fitted curve to the coefficient in equation (16). Assuming that the HIV infection began in 1979, find an estimate for the cumulative number of HIV infections in 1992.

References

Anderson, Roy and Robert May (May 1992), "Understanding the AIDS Pandemic," *Scientific American*, pages 58-66.

Bailey, N. T. J. (1975), *The Mathematical Theory of Infectious Diseases,* Griffin, New York.

Bremermann, H. J. and R.J. Anderson (1990), "Mathematical Models of HIV Infection," *Journal of Acquired Immune Deficiency Syndromes* 3: 1129-1134.

Castillo-Chavez, C. (Ed.) (1989), *Mathematical and Statistical Approaches to AIDS Epidemiology,* Springer Verlag.

Colgate, S., E. Stanley, J. Hyman, S. Layne and C. Qualls (1989), "Risk behavior-based model of the cubic growth of acquired immunodeficiency syndrome in the United States," *Proceedings of the National Academy of Sciences* 86.

Hethcote H.W. and J.A. Yorke (1984), *Gonorrhea Transmission Dynamics and Control,* Springer Verlag.

Marks, Geoffrey and William Beatty (1976), *Epidemics*, Charles Schribner's Sons, New York.

Answers to Exercises

1. 2.06 weeks. $R_0e^{-k(t+t_1)} = 2R_0e^{-kt} \Longrightarrow t_1 = \dfrac{\ln 2}{k}$. Here t_1 is the doubling time and is independent of t.

2. It might help to substitute $x = \dfrac{R(t)}{N}$.

4. The epidemic curve is the graph of $R'(t) = \dfrac{R_0kN^2(N-R_0)e^{-kNt}}{[R_0+(N-R_0)e^{-kNt}]^2}$. It reaches its maximum at $t = \dfrac{1}{kN}\ln\dfrac{N-R_0}{R_0}$.

5. At 1 week exponential gives 33, logistic gives 33. At 4 weeks exponential gives 1215, logistic gives 979.

6. $\lim_{t\to\infty} R(t) = 0$. The infection dies out.

7. Look at the Cumulative column. Notice, for example, that the numbers go from 685 to 1374 in one-half year. The doubling time is about one-half year. Later, numbers go from 66,290 to 135,614 in two years. The doubling time is increasing. It does not seem to be exponential growth.

8. Some major assumptions:
- L is constant over the interval 2 to 18.
- $N(r)$ is proportional to r^{-3}.
- The growth of infection in a risk group is exponential.
- The growth rate in a risk group is proportional to r.
- The quantity $N(r)/H_r(0)$ is constant with respect to r.
- The infection spreads as a wave so that we can estimate the total infected by adding up the groups that are currently saturated.

9. Substitute $u = t - \tau$ and use the Fundamental Theorem of Calculus. If $A'(t) = 0.0625[K(t-2)^2 - K(t-18)^2] = 0.0625K(32t - 320)$, then $A(t)$ is quadratic.

10. $A(t) = \frac{K}{12\cdot 128}[t^4 - 8t^3 + 24t^2 - 32t + 16]$. Hence it is a quartic function of time.

11. If $A(t)$ is exponential, then $A'(t)$ is exponential and therefore $H(t)$ is also. This means the integral in (12) must be an exponential function of the form $Ce^{\alpha/r}$. From this it follows that $N(r)$ has the form $\frac{C}{r^2}e^{\alpha/r}$.

12. $187 = \frac{0.0625K}{3}$ gives $K = 8976$. For the year 1992, $t = 13$ and $H(13) \approx K(13)^2 = 1,516,944$.

SPEEDY SORTING

Author: Steven Janke, Colorado College, Colorado Springs, CO 80903

Calculus Prerequisites: Limits at infinity, L'Hôpital's rule, integral of logarithm.

Area of Application: Computer science

New Mathematics Developed: Order of a function, approximation of $N!$

Note: Although not necessary, demonstration of sorting on a computer would add to the material.

The Problem: Sorting Large Lists

Computers spend a large portion of their time sorting data. Business computers keep mailing lists in order, college computers keep alumni records in order, and research computers often order data when compiling summary statistics. Although the task of putting 1000 test scores in numerical order may not be complicated, it certainly could be time consuming if we had to do it by hand. Even a computer takes some time. The following table lists the amount of time it took a microcomputer to put lists of random numbers in order:

List Size	Time (seconds)
1000	2.69
2000	11.78
4000	47.51
8000	190.70

Table 1. Sorting Times
(by Selection Sort on a 386 microprocessor running at 20 mHz).

One interesting feature of the data is that the time does not simply double when the list size doubles—it increases about four-fold. A little arithmetic shows that the time to sort 64,000 numbers would be over 3 hours! Such a list size is not uncommon for computers at the National Weather Service, the Center for Disease Control, or large insurance companies. Hence it might pay to look closer at the sorting problem to see if we can reduce the time and perhaps discover something about optimal sorting algorithms.

Selection Sort

To sort numbers, the computer must have the numbers in memory. It is convenient to think of memory locations as labeled boxes sitting in a row. A common way of denoting the numbers in a list is to call the number in the first box LIST[1], the number in the second box LIST[2], etc. The computer can compare the contents of two boxes and swap the contents of two boxes. The task now is to put the numbers in increasing order by using only these two operations: comparisons and swaps.

One intuitive way for putting the numbers in order is first to scan the list and put the smallest number in the first position. Next, disregard the first number and consider the rest. Find the smallest of these and put it in the second position. Continue in this fashion until reaching the end of the list. This method for sorting a list of numbers is called the Selection Sort.

Let us look at the details of this sorting method for a list of size ten. The numbers are denoted LIST[1] through LIST[10]. Compare LIST[1] to LIST[2]. If LIST[1] is larger, swap LIST[1] and LIST[2]; otherwise leave the numbers alone. If a swap was made, the numbers in boxes 1 and 2 have changed places so now LIST[1] is the number that was previously LIST[2], and LIST[2] is the number that was previously LIST[1]. Next compare LIST[1] to LIST[3] and again swap if LIST[1] is the larger. Continue in this way until LIST[1] is compared to LIST[10] and swapped if necessary. After this pass through the list, the smallest number in the list is LIST[1] and it is in its correct position at the left of the list.

Now begin with LIST[2] and compare it to LIST[3] through LIST[10], swapping when necessary. After this pass through the list, LIST[2] is the second smallest number in the entire list. Selection Sort continues by comparing LIST[3] to LIST[4] through LIST[10]. Each number is considered in turn until LIST[9] is compared to LIST[10]. After this stage, the list is in order. Figure 1 shows the results of three passes of Selection Sort on an original random list.

Original list:	[7]	[3]	[8]	[1]	[4]	[6]	[5]	[10]	[2]	[9]
First pass:	1	[7]	[8]	[3]	[4]	[6]	[5]	[10]	[2]	[9]
Second pass:	1	2	[8]	[7]	[4]	[6]	[5]	[10]	[3]	[9]
Third pass:	1	2	3	[8]	[7]	[6]	[5]	[10]	[4]	[9]

Figure 1. Selection Sort (three passes)

Let's summarize Selection Sort. Suppose the list has size N. Then when using Selection Sort, LIST[i] is compared to each of the numbers LIST[$i+1$] through LIST[N]. Two numbers in a comparison are swapped if the first is larger than the second. Selection Sort can be presented conveniently in the following pseudocode:

For $i = 1$ to $N - 1$ do
  For $j = i + 1$ to N do
    Compare LIST[i] to LIST[j] and swap if LIST[i] is larger.

Finally, notice that the Selection Sort is not specific to numbers. Clearly, if we stored words in the boxes and compared the contents of two boxes using alphabetic ordering, we could put a list of words in alphabetical order.

Exercise

1. Refer to Figure 1 and show the result after the fourth and fifth passes.

Analysis of the Selection Sort

We will try to predict how long it will take a computer to perform Selection Sort on a list of a given size. To begin with, we will simplify the problem slightly by assuming that the computer spends time only on comparisons and swaps. There are other operations the computer must perform, such as reading in the list, printing it out, and keeping track of where it is in the list, but the dominant parts of the selection algorithm are the comparisons and swaps. If we know how long the computer takes to perform one comparison and one swap, then to find out the total time for the sorting process we count the total comparisons and the total swaps. Multiplication will give us total time.

In Selection Sort, the first pass through the list compared LIST[1] to every other number. Since we had ten numbers, there were 9 comparisons on the first pass. On the second pass, LIST[2] was compared with each number to the right of it giving a total of 8 comparisons. Each successive pass makes one less comparison. The total number of comparisons to perform Selection Sort on 10 numbers is therefore $9 + 8 + 7 + 6 + 5 + 4 + 3 + 2 + 1 = 45$.

It isn't hard to generalize to lists of arbitrary size. If the list has N numbers, the first pass will need N-1 comparisons, the second pass will need N-2 comparisons, and the last pass will need 1 comparison. The total number of comparisons is

$$\sum_{i=1}^{N-1} i = \frac{N(N-1)}{2}. \tag{1}$$

We also want to count the number of swaps necessary to perform Selection Sort. However, the number of swaps depends on the particular list. The worst case occurs when the list is in reverse order; the best case occurs when the list is already in order. What would seem appropriate here is the <u>average</u> number of swaps. This takes a little work to calculate and can be found in Knuth (1973). One thing we can say for sure is that there are never more swaps than there are comparisons. Consequently, we can get a good idea of how our algorithm performs on various lists by counting the number of comparisons. The result for Selection Sort given here is that it always takes $\frac{N(N-1)}{2}$ comparisons.

Exercise

2. Show that formula (1) is correct. (Hint: Add the first and last numbers, then add the second and second from last, etc.)

Insertion Sort

We will look briefly at one more simple sorting algorithm. This one, called the Insertion Sort, orders a list in the same way that we usually order a hand of playing cards. Imagine that someone deals you a hand of playing cards. You pick up the cards one at a time and insert them in the fan of cards held in your other hand. After you have picked them all up, you have an ordered fan.

The same technique works in a computer. A row of memory boxes containing numbers represents the dealt hand of cards. The number at the left, LIST[1], can be thought of as an ordered list of size one. This starts the "fan" of ordered numbers. Next "pick up" LIST[2] and compare it to LIST[1]. Swap if necessary. Now there is an ordered list of size two at the left of the row. Consider LIST[3] and compare it to LIST[2] swapping if they are out of order. If it was not necessary to swap, then the numbers LIST[1], LIST[2], LIST[3] are in order. If it was necessary to swap, then LIST[2] must be compared to LIST[1] and swapped if necessary. In either case, there is now an ordered list of size three at the left of the row. The algorithm continues by considering LIST[4] and "inserting" it in the ordered list at the left by comparing and perhaps swapping. Remember that if a swap is not necessary, then the partial list already is in order. Figure 2 shows the first stages of Insertion Sort as it operates on a random list.

Original list:	5	[3]	[8]	[4]	[10]	[6]	[1]	[9]	[7]	[2]
First Stage:	3	5	[8]	[4]	[10]	[6]	[1]	[9]	[7]	[2]
Second Stage:	3	5	8	[4]	[10]	[6]	[1]	[9]	[7]	[2]
Third Stage:	3	4	5	8	[10]	[6]	[1]	[9]	[7]	[2]
Fourth Stage:	3	4	5	8	10	[6]	[1]	[9]	[7]	[2]

Figure 2. Insertion Sort (first stages)

Here is the pseudocode for Insertion Sort on a list of size N:

```
For i = 2 to N do
    Set j = i.
    While j > 0 and LIST[j]<LIST[j − 1] do
        Swap LIST[j] and LIST[j − 1]. Subtract one from j.
```

Analysis of the Insertion Sort

One important feature of Insertion Sort is that sometimes further comparisons are avoided. If a swap is not made after comparing two numbers, then Insertion Sort stops further comparisons and goes on to consider the next unexamined number. To compare Insertion Sort to Selection Sort, we will calculate the average number of comparisons for Insertion Sort.

To do this, suppose that we are in the middle of an insertion sort, J numbers are in order at the left of the list, and a new number is about to be inserted. The new number could fit between any of the J numbers or it could fit at the beginning or end of the list. There are $J+1$ possible positions for the new number. If we assume that each position is equally likely, then the probability that the number will be in any given position is $\frac{1}{J+1}$.

If the new number comes at the beginning of the list, it must be compared with all the J numbers before it is put where it belongs. That is, J comparisons are needed. The new number must also be compared with all J numbers in order to discover that it belongs between the first and second. If it comes between the second and third, there will be only $J-1$ comparisons. Between the third and fourth requires only $J-2$ comparisons. If the new number comes after all the numbers in the list, then only one comparison is necessary. To calculate the average number of comparisons, we multiply the number of comparisons necessary to put it in a particular position by the probability $\frac{1}{J+1}$ that it will be in that position, and then sum:

$$\begin{aligned}\text{Average comparisons, } J\text{th stage} &= \frac{1}{J+1}(J + J + (J-1) + (J-2) + \cdots + 2 + 1) \\ &= \frac{J}{J+1} + \frac{1}{J+1}\sum_{i=1}^{J} i \qquad (2) \\ &= \frac{J}{J+1} + \frac{J}{2}.\end{aligned}$$

To finish the calculation for the average number of comparisons in Insertion Sort, we sum over all values of J:

$$\begin{aligned}\text{Average comparisons} &= \sum_{J=1}^{N-1}\left(\frac{J}{2} + \frac{J}{J+1}\right) \qquad (3) \\ &= \frac{N(N-1)}{4} + \sum_{J=1}^{N-1}\frac{J}{J+1}.\end{aligned}$$

It is difficult to calculate the sum $\sum_{J=1}^{N-1}\frac{J}{J+1}$ directly, but we can get good upper and lower bounds for it by using calculus. First simplify the sum by noting that

$$\sum_{J=1}^{N-1}\frac{J}{J+1} = \sum_{J=1}^{N-1}\left(1 - \frac{1}{J+1}\right) = N - 1 - \sum_{i=2}^{N}\frac{1}{i}. \qquad (4)$$

Hence we need to find upper and lower bounds for the sum of $1/i$ as i runs from 2 to N. In Figure 3, notice that this sum is the sum of rectangles under the curve $y = 1/x$ from 1 to N. Hence an upper bound for the sum is $\int_1^N \frac{1}{x} dx = \ln N$. To find a lower bound, again notice from the figure that the dotted rectangles include the area under the curve, so the sum is greater than $\int_2^{N+1} \frac{1}{x} dx = \ln(N+1) - \ln 2$. Thus

$$N - 1 - \ln N < \sum_{J=1}^{N-1} \frac{J}{J+1} < N - 1 - \ln(N+1) + \ln 2. \tag{5}$$

These bounds will be useful in the next section.

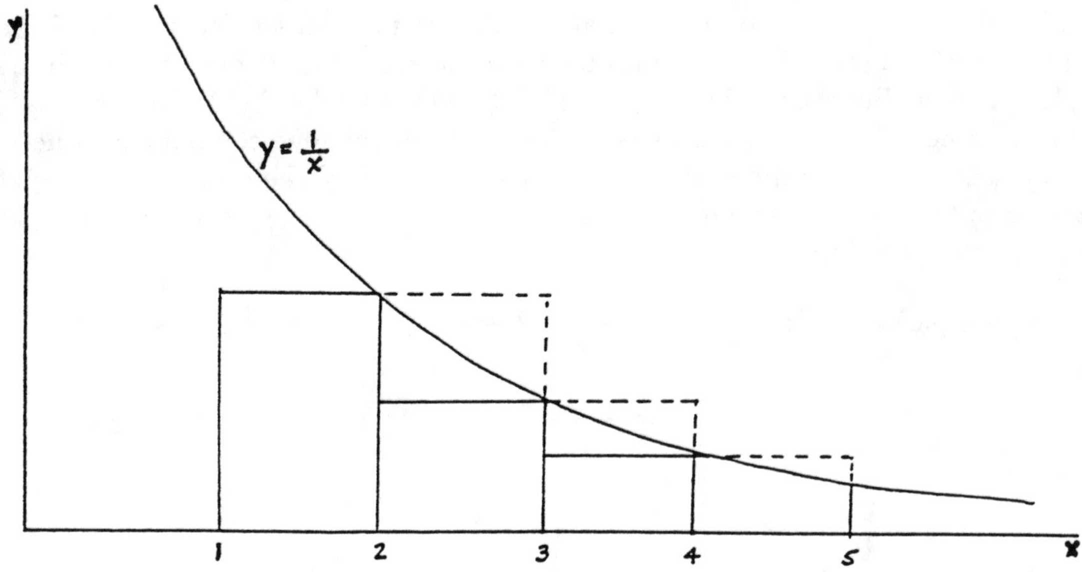

Figure 3. Estimating $\sum_{i=2}^{N} \frac{1}{i}$.

In formula (3) for the average number of comparisons in Insertion Sort, the first term is the dominant one. For example, if $N = 1000$, the first term is 249,750 and the second term is about 992.5. Therefore, when we compare (3) to (1), it appears that the average number of comparisons for Insertion Sort is about half the average number for Selection Sort. This suggests that Insertion Sort might run faster, and indeed it does, as Table 2 shows.

List Size	Time (Seconds)
1000	1.73
2000	7.46
4000	29.98
8000	73.47

Table 2. Insertion Sort Times

For both Table 1 and Table 2, the sorting algorithms were run on two random lists of the given size and the average time was calculated. Insertion Sort is about twice as fast as Selection Sort.

Exercises

3. Figure 2 shows a list of ten numbers. How many comparisons does it take Insertion Sort to put this list in order? What is the average number of comparisons for Insertion Sort on a list this size?

4. Use integrals to find upper and lower bounds for the following sums: $\sum_{i=1}^{5} i^3$ and $\sum_{i=1}^{4} 3^i$. Compare your bounds to the actual sums.

The Order of an Algorithm

We have found functions that give the average number of comparisons necessary for Selection Sort and Insertion Sort. If N is the size of the list, they are:

$$\text{Selection Sort: } C_S(N) = \frac{1}{2}N^2 - \frac{1}{2}N$$

$$\text{Insertion Sort: } C_I(N) = \frac{1}{4}N^2 - \frac{1}{4}N + \sum_{J=1}^{N-1} \frac{J}{J+1}.$$

We will call functions that count the average number of comparisons for a sort *C-functions*. We have seen that the C-function for Insertion Sort is about half the C-function for Selection Sort. The next goal is to determine some systematic way for comparing two C-functions.

For small lists, the time for sorting—by any algorithm—is rather small, so how C-functions compare is not of practical importance. However, for large lists the sorting time becomes significant and hence the differences between algorithms can be crucial. Our interest in large N suggests that we consider the limits of C-functions as N approaches infinity. Hence to compare two C-functions, we will consider their ratio, and look at the limit of the ratio as N approaches infinity.

Suppose, for example, that we have two other sorting algorithms, called A and B, with C-functions $C_A(N) = 3N$ and $C_B(N) = \frac{1}{4}N^2$. We would compare them to $C_S(N)$

by calculating

$$\lim_{N\to\infty} \frac{C_S(N)}{C_A(N)} = \lim_{N\to\infty} \frac{\frac{1}{2}(N^2 - N)}{3N} = \infty$$

$$\lim_{N\to\infty} \frac{C_S(N)}{C_B(N)} = \lim_{N\to\infty} \frac{\frac{1}{2}(N^2 - N)}{\frac{1}{4}N^2} = 2.$$

Looking at the first limit, the ratio of the Selection Sort C-function to the C-function for algorithm A gets arbitrarily large as N increases. This means that the two functions grow at different rates, and in fact C_S grows much faster than C_A. Consequently, Selection Sort would run much slower on large lists. The second limit shows that the function C_S is very close to twice C_B as N gets large. This indicates that the two C-functions grow at a comparable rate, although one is always smaller than the other.

The limit of the ratio is a convenient way of comparing the growth of two functions. However, since the limit may not always exist, we will use definitions that do not involve limits.

Definition. *Let f and g be two positive functions. Function f is said to be "big-oh" of g if there exist numbers c and N_0 with $c > 0$ such that $f(N) \leq cg(N)$ for all $N \geq N_0$. We use the notation $f = O(g)$.*

In other words, if $f = \text{O}(g)$, then f grows no faster than g. As examples, consider the functions C_A, C_B, and C_S. Since $3N < \frac{1}{2}(N^2 - N)$ for $N \geq 8$ (see Exercise 5), we have $C_A = \text{O}(C_S)$. Further, since $\frac{1}{2}(N^2 - N) < 2 \cdot \frac{1}{4}N^2$, it is also true than $C_S = \text{O}(C_B)$.

In order to indicate that two functions grow at the same rate, we introduce one more definition.

Definition. *Functions f and g have the same order if $f = O(g)$ and $g = O(f)$. We often say that g has order f.*

Above we showed that $C_S = \text{O}(C_B)$. Now notice that if $N > 4$ then $\frac{1}{4}N^2 < \frac{1}{2}(N^2 - N)$. Therefore, $C_B = \text{O}(C_S)$, and hence C_S and C_B have the same order.

When the limit of the ratio of two C-functions does exist, we can use it to establish the order of functions:

Proposition. *If $\lim_{N\to\infty} \frac{f(N)}{g(N)} = c$ and $0 < c < \infty$, then f and g have the same order. If $c = 0$ or $c = \infty$, then f and g have different orders.*

Proof: If $0 < c < \infty$, then there exist k_1 and k_2 with $0 < k_1 < c < k_2 < \infty$. Since the limit is c, there exists an N_0 such that for $N \geq N_0$

$$k_1 < \frac{f(N)}{g(N)} < k_2,$$

so that $k_1 g(N) < f(N) < k_2 g(N)$. This shows that f and g have the same order.

If $c = 0$, then for any $\epsilon > 0$, there is an N_ϵ such that for $N > N_\epsilon$

$$\frac{f(N)}{g(N)} < \epsilon.$$

Therefore, $g(N) > \frac{1}{\epsilon}f(N)$. Since g is eventually larger than any constant times f, $g \neq O(f)$. Hence f and g do not have the same order.

Finally, if $c = \infty$, then $\lim_{N\to\infty} \frac{g(N)}{f(N)} = 0$ and the previous case applies. Q.E.D.

The proposition shows that C_A and C_S have different orders and consequently, based on the number of comparisons, algorithm A would be far better than Selection Sort.

For another example, consider the two functions $100N^{1.5}$ and $5N^2 - 4N$. Since the limit as $N \to \infty$ of the ratio of the first function to the second is 0, the two functions have different orders. The second function grows faster than the first so we say it has higher order and represents a slower algorithm.

Sometimes finding the limit of a ratio is not straightforward. Suppose the two C-functions are $\ln N$ and N. Then in order to find $\lim_{N\to\infty} \frac{\ln N}{N}$ we can recall L'Hôpital's rule from calculus.

L'Hôpital's Rule. *If* $\lim_{N\to\infty} f(N) = \lim_{N\to\infty} g(N) = \infty$, *and if* f' *and* g' *exist, then* $\lim_{N\to\infty} \frac{f(N)}{g(N)} = \lim_{N\to\infty} \frac{f'(N)}{g'(N)}$.

Applying the rule to the ratio of $\ln N$ and N gives

$$\lim_{N\to\infty} \frac{\ln N}{N} = \lim_{N\to\infty} \frac{1/N}{1} = 0.$$

Thus the two functions have different orders.

Now we are in a position to compare the C-functions for the Selection Sort and the Insertion Sort. From (3) and (5),

$$\frac{1}{4}(N^2 - N) + N - 1 - \ln N < C_I(N) < \frac{1}{4}(N^2 - N) + N - 1 - \ln(N+1) + \ln 2. \qquad (6)$$

Consider first the function forming the lower bound. Notice

$$\begin{aligned}
\lim_{N\to\infty} \frac{C_I(N)}{C_S(N)} &> \lim_{N\to\infty} \frac{\frac{1}{4}(N^2 - N) + N - 1 - \ln N}{\frac{1}{2}(N^2 - N)} \\
&= \lim_{N\to\infty} \left(\frac{1}{2} + \frac{2(N-1)}{N^2 - N} - \frac{2\ln N}{N^2 - N}\right) \qquad (7) \\
&= \lim_{N\to\infty} \left(\frac{1}{2} + \frac{2}{N} - \frac{2/N}{2N-1}\right) = \frac{1}{2}.
\end{aligned}$$

Similarly, the ratio of the upper bound to $C_S(N)$ also tends to $1/2$ as $N \to \infty$. Hence, the limit of the ratio of C_I to C_S is $1/2$ and the two functions have the same order.

Exercises

5. Show that if $N \geq 8$ then $C_A(N) < C_S(N)$.

6. Show that C_I and C_S have the same order as $f(N) = N^2$.

7. Show that $f(N) = 10N^5 - 3N^3 + N^2 - 100$ has the same order as $g(N) = N^5 - 2N$.

8. List the following functions in increasing order. Also indicate if two functions have the same order.

$N^2 - N^{1.5}$	$\ln N^2$	$(\ln N)^2$
$N \ln N$	2^N	$N + \ln 3N$
$\log_2 N$	$N^{0.01}$	N^{100}

Optimal Order for Sorting

Insertion Sort and Selection Sort both have order N^2 for the average number of comparisons. Are there other algorithms that do better? Is order N^2 the best possible? In fact, there are several algorithms that are better, and it is even possible to give a lower bound on the order of the best possible C-furction.

To estimate the C-function for an optimal sorting algorithm, we will consider what kind of work a sorting algorithm must do. If we have a list of N numbers, in how many different ways can they be ordered? The first element can be any one of the N numbers. Once the first element has been determined, there are $N-1$ choices for the second element, $N-2$ choices for the third element, etc. To count the total number of orderings, multiply all these choices together. The total number is $N(N-1)(N-2)(N-3)\cdots 1 = N!$ Any sorting algorithm must be able to distinquish the $N!$ different orderings of N numbers.

One way to diagram what a sorting algorithm does is to draw a *comparison tree.* The diagram in Figure 4 shows how Selection Sort compares elements in a list of size three.

To read the diagram, let the numbers in the list be a_1, a_2, and a_3. The circle at the top of the tree indicates that Selection Sort first compares a_1 to a_2. If $a_1 < a_2$, follow the left branch of the tree, otherwise follow the right branch. For example, if $a_1 < a_2$, Selection Sort next compares a_1 to a_3. On the other hand if $a_1 > a_2$, then Selection Sort compares a_2 to a_3. The rectangles at the bottom of the tree are called "leaves" (despite the fact that they are at the bottom!), and indicate the correct increasing order for the three numbers. Since there are $3! = 6$ ways to order three numbers, there are six leaves in this comparison tree.

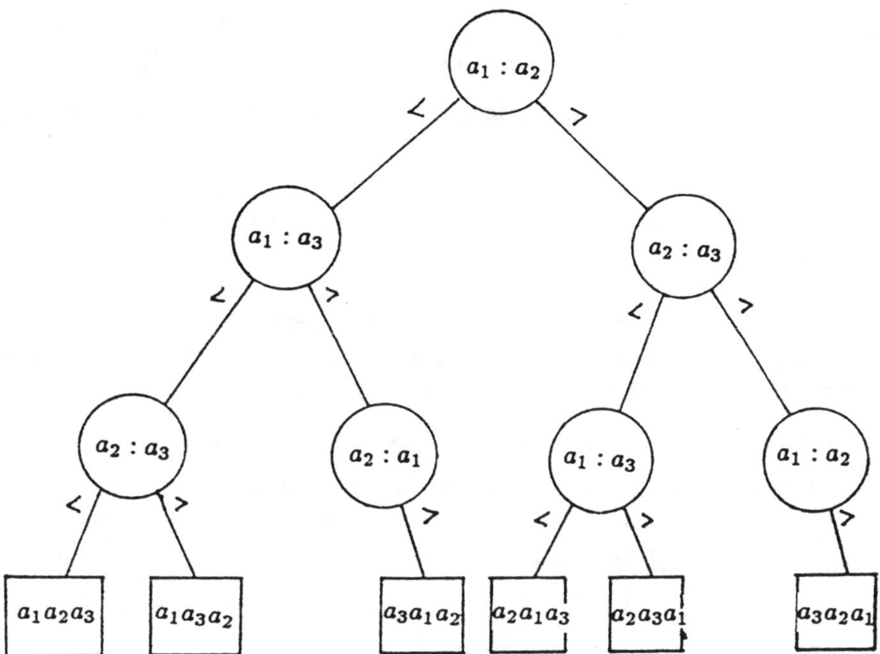

Figure 4. Comparison Tree.

The circle labeled $a_2 : a_1$ has only one branch, leading to the ordering $a_3a_1a_2$. There is no other possibility because at this stage in the tree we know that a_2 is larger than a_1, since these numbers were already compared at the root node of the tree. Thus the circle $a_2 : a_1$ is redundant, as is the circle $a_1 : a_2$ at the right edge of the tree. Since these redundancies represent inefficiencies in Selection Sort, we can imagine a slightly better sort where there are no redundant comparisons. For a best possible sorting algorithm, we can assume that each circle in the comparison tree has two branches leading from it; that is, there are no redundancies.

Comparison trees have a number of levels. In Figure 4, the tree has a single root node at the top and three levels of nodes underneath. We say that this tree has ***depth*** three. Any path traced from the root to a leaf indicates a possible set of comparisons that Selection Sort might do when ordering a list of numbers. The number of comparisons in the worst case is the number of circles on the longest path, and this is the depth of the comparison tree.

If we could show that any comparison tree for a list of size N must have a depth of at least K_N, then we would know that even the best possible algorithm would require K_N comparisons in the worst case. To find such a lower bound, we need the following result.

Proposition. *A comparison tree which has M leaves has depth at least $\lceil \log_2 M \rceil$.*

(The symbols $\lceil$ and $\rceil$ indicate the least integer greater than or equal to $\log_2 M$. This is read "ceiling of $\log_2 M$.")

Proof: A comparison tree has two branches out of every comparison node and none out of the leaf nodes. If the tree has depth d, then the tree has at most 2^d leaves. Hence $M \leq 2^d$.

By taking logarithms to the base 2, we get $\log_2 M \leq d$. Since d is an integer, d must be at least as large as the least integer greater than or equal to $\log_2 M$. Q.E.D.

Now we can give a lower bound on the number of comparisons that any algorithm must do.

Proposition. *On a list of size N, any sorting algorithm must do at least $\lceil \log_2 N! \rceil$ comparisons in the worst case.*

Proof: If the list has size N, there are $N!$ possible orderings and hence the comparison tree has at least $N!$ leaves. By the previous proposition, the depth of the tree must be at least $\lceil \log_2 N! \rceil$. Q.E.D.

This is a remarkable result. Without knowing very many sorting algorithms, we have been able to prove that *any* sorting algorithm that sorts data by comparing elements must do at least $\lceil \log_2 N! \rceil$ comparisons to sort some list of size N. Note that we have not proved that there is any algorithm that is this good.

The function $f(N) = \lceil \log_2 N! \rceil$ is a lower bound for the worst case in sorting a list. In order to find a lower bound for the average case (hence a lower bound for the best possible C-function), we need to go back to the comparison tree and find the average length of a path from the root to a leaf. This takes a little more effort, but it can be done.

Proposition. *The best possible C-function must satisfy $C(N) \geq \lfloor \log_2 N! \rfloor$. In other words, the average number of comparisons done by any sorting algorithm on lists of size N is at least $\lfloor \log_2 N! \rfloor$.*

(The symbols $\lfloor$ and $\rfloor$ indicate the greatest integer less than or equal to $\log_2 N!$. This is read "floor of $\log_2 N!$".)

Proof: See Baase (1988).

The lower bound for the average case differs from the lower bound for the worst case, but as you might suspect, they have the same order. In fact, since $\lfloor \log_2 N! \rfloor \leq \log_2 N! \leq \lceil \log_2 N! \rceil$ and the two outer functions differ by at most 1, all three functions have the same order. Is this optimal order close to the order for the Selection and Insertion Sorts? In other words, does $\log_2 N!$ have the same order as N^2? It turns out that it doesn't.

Proposition. *The function $f(N) = \log_2 N!$ has order $N \ln N$.*

Proof: There is one technical detail we must deal with first. The function $f(N)$ involves a logarithm to the base 2. Recall that $\log_2 x = \ln x / \ln 2$. Logarithms to the base 2 are just constant multiples of natural logarithms. Therefore, $\log_2 N!$ and $\ln N!$ have the same order, and we can work with $\ln N!$

Now, $\ln N! = \ln N(N-1)(N-2)\cdots 1 = \sum_{i=1}^{N} \ln i$. Once again we can use an integral to estimate the sum. From Figure 5, we can see that

$$\int_1^N \ln x \, dx \leq \sum_{i=1}^{N} \ln i \leq \int_1^{N+1} \ln x \, dx.$$

Since $\int \ln x\, dx = x \ln x - x + C$, we get that

$$N \ln N - N + 1 \leq \sum_{i=1}^{N} \ln i \leq (N+1)\ln(N+1) - (N+1) + 1. \tag{8}$$

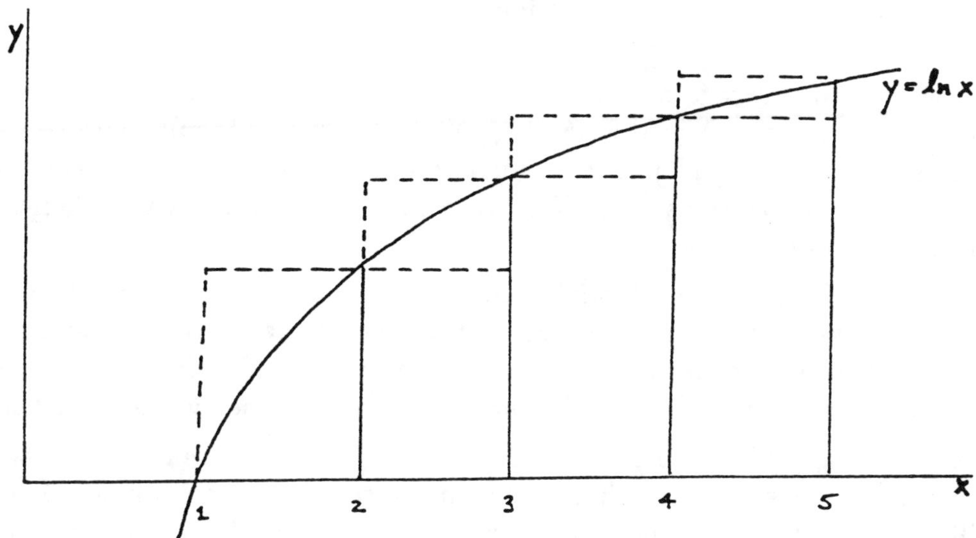

Figure 5. Estimating $\sum_{i=1}^{N} \ln i$.

The sum is bounded above and below by two functions which both have order $N \ln N$, since

$$\lim_{N\to\infty} \frac{N \ln N - N + 1}{N \ln N} = 1,$$

$$\lim_{N\to\infty} \frac{(N+1)\ln(N+1) - N}{N \ln N} = 1.$$

(L'Hôpital's Rule helps in establishing the second limit.)

Hence $\lim_{N\to\infty} \dfrac{\sum_{i=1}^{N} \ln i}{N \ln N} = 1$ since this limit is squeezed between the previous two. This shows that the sum has order $N \ln N$. Q.E.D.

Exercises

9. Verify the inequalities involving $\sum_{i=1}^{N} \ln i$, and then do the integration by parts to verify the upper and lower bounds.

10. Use L'Hôpital's rule to establish $\lim_{N\to\infty} \dfrac{(N+1)\ln(N+1) - N}{N \ln N} = 1.$

11. In the proof of the last proposition, we trapped $\ln N! = \sum_{i=1}^{N} \ln i$ between two functions. We could use the average of the upper and lower functions to estimate $\ln N!$, and then use $N! = e^{\ln N!}$ to estimate $N!$. Use this technique to estimate 10! and compare it to the correct value. (By being a little more careful about the estimate, we can derive what is called Stirling's Approximation.)

Quicksort

We know that the average number of comparisons done by the best possible sorting algorithm sorting a list of size N is at least $\lfloor \log_2 N! \rfloor$. It may be difficult to find an algorithm whose C-function equals $\lfloor \log_2 N! \rfloor$, but it is not too hard to discover algorithms whose C-functions have the same order as $\lfloor \log_2 N! \rfloor$. One such algorithm is called Quicksort, and we will now explain how it works.

A good idea in designing an algorithm to solve a problem is to break the problem up into smaller pieces. When faced with a large list to sort, it may prove efficient to work on the list until we have two smaller lists where every number in the first list is smaller than every number in the second list. Then the problem is to sort the two smaller lists.

This is the idea behind Quicksort. Split a list into two smaller ones with every element in the first list less than every element in the second. Then split each smaller list into two still smaller lists. Continue in this way until the small lists have only one or two elements, and we can sort them with at most one comparison each.

To split a list into two smaller ones, Quicksort starts by selecting one of the numbers as a *pivot*. Then it compares every other element in the list to the pivot in order to decide whether it comes before pivot or after it. Thus it requires $N-1$ comparisons to split a list of size N; one number is the pivot, and all others are compared to it.

Counting the average number of comparisons that Quicksort needs to sort a list of size N is a little complicated, but we can get a good idea how the analysis proceeds by simplifying the problem. First, assume that $N = 2^k - 1$. Also assume that every time we split the list the pivot ends up right in the middle. For example, if $N = 7$, then after the first split there will be two smaller lists of size 3 and the pivot will be in the middle. Each smaller list will be split into two lists of size 1 with the pivot in the middle.

Let $C_Q(N)$ be the number of comparisons necessary to sort a list of size N. By our assumptions, after the first split the two smaller lists each have size $(N-1)/2$, and therefore

$$C_Q(N) = 2C_Q(\frac{N-1}{2}) + N - 1, \tag{9}$$

since it takes $N-1$ comparisons to split the list. Equation (9) is called a *recurrence relation* since the function C_Q is defined in terms of itself. Also note that since a list of size one requires no comparisons to order, $C_Q(1) = 0$.

Since we have assumed that $N = 2^k - 1$, it is easy to use the recurrence relation (9) to calculate $C_Q(N)$:

$$
\begin{aligned}
C_Q(N) = C_Q(2^k - 1) &= 2C_Q(2^{k-1} - 1) + 2^k - 2 \\
&= 2(2C_Q(2^{k-2} - 1) + 2^{k-1} - 2) + 2^k - 2 \\
&= 2^2 C_Q(2^{k-2} - 1) + 2 \cdot 2^k - 2 - 4 \\
&= \cdots \\
&= 2^{k-1} C_Q(1) + (k-1) \cdot 2^k - \sum_{i=1}^{k-1} 2^i \\
&= (k-1) \cdot 2^k - (2^k - 2).
\end{aligned} \tag{10}
$$

Now $k = \log_2(N+1)$, so $C_Q(N) = (N+1)\log_2(N+1) - 2N$, which has has order $N \ln N$. We have derived the function C_Q only in a special case, but it turns out that this special case is indicative of the average case order of Quicksort.

Proposition. *The average case order of Quicksort is $N \ln N$.*

Proof: See (Baase, 1988).

The data in Table 3 shows that Quicksort lives up to its name. The times in the table were obtained by running Quicksort on the same lists used in compiling Tables 1 and 2.

List Size	Selection Sort	Insertion Sort	Quicksort
1000	2.69	1.73	0.11
2000	11.78	7.46	0.22
4000	47.51	29.98	0.44
8000	190.70	73.47	0.96

Table 3. Selection Sort, Insertion Sort, and Quicksort Times (in seconds)

Exercise

12. Show that $C_Q(N)$ derived above has order $N \ln N$.

Calculus in Computer Science

Computer science is a combination of designing algorithms and analyzing how well they work. Sorting algorithms offer a good example of how calculus is an important tool in determining how fast an algorithm will run.

In the analysis of algorithms, calculus is particularly useful in comparing functions (e.g. the definition of order of a function), in estimating sums (e.g. Stirling's approximation in Exercise 11), and in solving recurrence relations (e.g. the relation derived for Quicksort). Knuth (1973) has a wealth of results that use these calculus methods.

References

Baase, Sara (1988), *Computer Algorithms*, Addison Wesley.

Knuth, Donald (1973), *The Art of Computer Programming*, Volume 3, Addison Wesley. This classic book is full of interesting results on sorting and searching. Many require advanced mathematics, but there is something for everyone.

Sedgewick, Robert (1988), *Algorithms*, Addison Wesley.

Answers to Exercises

1. Fourth pass: 1 2 3 4 8 7 6 10 5 9. Fifth pass: 1 2 3 4 5 8 7 10 6 9
2. If $N-1$ is even, there are $\frac{N-1}{2}$ pairs of numbers. Pair the first and last, then second and next to last, etc. Each pair sums to N. Hence the total sum is $\frac{N-1}{2}N$. If $N-1$ is odd, then there are $\frac{N-2}{2}$ pairs that sum to N and one extra number in the middle equal to $N/2$. Hence the total sum is $\frac{N-2}{2}N + \frac{N}{2} = \frac{N-1}{2}N$. (This sum is also easily established by induction.)
3. Comparisons for list: 30 comparisons. Average for 10 numbers: ≈29.57 comparisons.
4. $\int_0^5 x^3 dx < \sum_{i=1}^{5} i^3 < \int_1^6 x^3 dx$ implies $156.25 < \sum_{i=1}^{5} i^3 < 323.75$. The actual sum is 225.

   $\int_0^4 3^x\, dx < \sum_{i=1}^{4} 3^i < \int_1^5 3^x\, dx$ implies $72.82 < \sum_{i=1}^{4} 3^i < 218.46$. The actual sum is 120.
5. One approach: $C_S(N) - C_A(N)$ is increasing as long as $N > 3.5$. (Check the first derivative.) Finally $C_S(7) - C_A(7) = 0$, therefore the difference is positive for $N \geq 8$.
6. $\lim_{N\to\infty} \frac{C_S(N)}{N^2} = \frac{1}{2}$. Therefore C_S has order N^2. C_I has the same order as C_S.
7. $$\lim_{N\to\infty} \frac{10N^5 - 3N^3 + N^2 - 100}{N^5 - 2N} = \lim_{N\to\infty} \frac{10 - \frac{3}{N^2} + \frac{1}{N^3} - \frac{100}{N^5}}{1 - \frac{2}{N^4}} = 10.$$
8. $\log_2 N \leftrightarrow \ln N^2 \rightarrow (\ln N)^2 \rightarrow N^{0.01} \rightarrow N + \ln 3N \rightarrow N \ln N \rightarrow N^2 - N^{1.5} \rightarrow N^{100} \rightarrow 2^N$.

11. $\ln N! \approx \frac{1}{2}[(N \ln N - N + 1) + ((N+1)\ln(N+1) - (N+1) + 1)]$. Therefore $\ln 10! \approx 15.20135$. This gives $10! \approx 3,998,180$. The actual value is 3,628,800.
12. Taking the limit of the ratio gives:

$$\begin{aligned}\lim_{N\to\infty} \frac{C_Q(N)}{N \ln N} &= \lim_{N\to\infty} \frac{(N+1)\log_2(N+1) - 2N}{N \ln N} \\ &= \lim_{N\to\infty} \frac{1}{\ln 2}\frac{(N+1)\ln(N+1)}{N \ln N} - \frac{2}{\ln N} \\ &= \lim_{N\to\infty} \frac{1}{\ln 2}[\frac{\ln(N+1)}{\ln N} + \frac{\ln(N+1)}{N \ln N}] \\ &= \lim_{N\to\infty} \frac{1}{\ln 2}\frac{N}{N+1} = \frac{1}{\ln 2}.\end{aligned}$$

HYDRO-TURBINE OPTIMIZATION

Author: adapted by Philip Straffin, Beloit College, Beloit, WI 53511

Source: Harry Bard, Great Northern Paper Company, Millinocket, Maine. This problem was originally prepared by an MAA Workshop Group whose members were Walter Brady, David Dimmock, Margaret Elliott, Ken Hamilton, Walter Jensen, Bruce Pyne and Dale Skrien, in a series edited by Jeanne Agnew and Marvin Keener, Oklahoma State University.

Area of application: industry, engineering.

Calculus needed: Lagrange multipliers.

Suggestions on use: a calculator is necessary for numerical work in the exercises.

The Problem: Getting the Most Power from Turbines

The Great Northern Paper Company in Millinocket, Maine, produces newsprint, computer paper, and many other kinds of paper goods. In order to ensure an adequate supply of affordable power, it also operates six hydro-electric generating stations on the Penobscot River. In the present problem we are concerned with the power station on the West Branch of the Penobscot River, which gets its water from a dam on Ripogenus Lake. A pipe sixteen feet in diameter and three-quarters of a mile long carries water from the dam to the power station, through an elevation drop of 170 feet. The rate at which water flows through the pipe varies, depending on conditions in the watershed.

Once at the power station, manually controlled valves and gates distribute the water to the station's three hydro-electric turbines. These turbines have known, and different, "power curves," which give the amount of electric power generated as a function of the water flow sent to the turbine. The problem, as presented to us by the power plant supervisor, is to devise a plan for distributing water among the turbines which will get the maximum energy production from the three turbines for any rate of water flow.

Modeling the Problem

Our plan will be to formulate the problem mathematically—build a *mathematical model* of the problem. We will then solve the mathematical problem, and translate our results into an operational plan for distributing water to the turbines. We start with some background.

Commercial electricity is produced by turbines which turn mechanical energy into electrical current. In some cases, coal, oil, gas or atomic fuel is used to make steam which runs the turbines. Hydro-electric power stations use the energy of falling water to turn the turbines. The energy comes both from the weight of the water and from the "head" on it, that is, the vertical distance through which the water falls.

The basic equation which relates water flow to energy production was published by Daniel Bernoulli in 1738, and is called *Bernoulli's equation.* It results from applying the principle of conservation of energy to the flow between the lake surface and the turbine. In our context, the equation states that

$$W = \gamma Q\eta(Z_h - Z_t - f), \tag{1}$$

where

$$\begin{aligned} W &= \text{power extracted by turbine (foot-pounds/second)} \\ \gamma &= \text{specific weight of water (pounds/foot}^3) \\ Q &= \text{flow rate of fluid (feet}^3\text{/second, abbreviated cfs)} \\ \eta &= \text{turbine efficiency, a function of } Q \\ Z_h &= \text{elevation of the lake surface (feet)} \\ Z_t &= \text{elevation of the turbine (feet)} \\ f &= \text{energy loss due to friction, a function of } Q. \end{aligned}$$

In our case, the difference between Z_h and Z_f is 170 feet. The main factor in f is the energy lost as water flows through the pipe. Engineers derived from experiment the estimate $f = 1.6 \cdot 10^{-6}\ Q_T^2$, where Q_T is the total water flow in cubic feet per second (cfs).

The efficiency η, which is a function of Q, differs for the three turbines. Experimental results suggested expressing $\gamma Q\eta$ as a quadratic polynomial in Q, for each turbine. Statistical curve fitting then gave the following equations for the power output of the three turbines:

$$\begin{aligned} KW_1 &= (-18.89 + 0.1277Q_1 - 0.408 \cdot 10^{-4}\ Q_1^2)(170 - 1.6 \cdot 10^{-6}\ Q_T^2) \quad 250 \le Q_1 \le 1110 \\ KW_2 &= (-24.51 + 0.1358Q_2 - 0.469 \cdot 10^{-4}\ Q_2^2)(170 - 1.6 \cdot 10^{-6}\ Q_T^2) \quad 250 \le Q_2 \le 1110 \\ KW_3 &= (-27.02 + 0.1380Q_3 - 0.384 \cdot 10^{-4}\ Q_3^2)(170 - 1.6 \cdot 10^{-6}\ Q_T^2) \quad 250 \le Q_3 \le 1225 \end{aligned} \tag{2}$$

where

$$\begin{aligned} Q_i &= \text{flow through turbine } i \text{ (cfs)} \\ KW_i &= \text{power generated by turbine } i \text{ (kilowatts)} \\ Q_T &= \text{total flow through the station (cfs).} \end{aligned}$$

The coefficients in the quadratic polynomials in (2) include a scaling factor to transform units of mechanical power into units of kilowatts. The bounds on the Q_i's represent the fact that the turbines cannot operate with a flow below 250 cfs, or above a maximum flow which is slightly higher for turbine 3 than for turbines 1 and 2.

If all three turbines are running, our problem of distributing water among the turbines to obtain the maximum energy production can now be formulated as a mathematical problem:

$$\text{Maximize} \qquad KW_1 + KW_2 + KW_3 \tag{3}$$

subject to

$$Q_1 + Q_2 + Q_3 = Q_T$$

$$\text{and} \quad 250 \le Q_1 \le 1110, \quad 250 \le Q_2 \le 1110, \quad 250 \le Q_3 \le 1225.$$

We must solve this problem for all feasible values of Q_T.

Lagrange Multipliers and a First Solution

The mathematical problem above is called a *constrained maximization* problem for a function of three variables. There is an elegant method for solving such problems, due to Joseph-Louis Lagrange (1736-1813) and now taught in several-variable calculus courses. I will present the method briefly here, and refer you to a calculus text for details.

Suppose we wish to maximize $f(x, y, z)$ subject to a constraint $g(x, y, z) = 0$, where f and g are differentiable functions. We observe that at a maximum (or minimum) point (x_0, y_0, z_0), the gradient vector $\nabla f(x_0, y_0, z_0)$ must be orthogonal to the surface $g(x, y, z) = 0$. Since the gradient $\nabla g(x_0, y_0, z_0)$ is also orthogonal to $g(x, y, z) = 0$ at (x_0, y_0, z_0), we see that we must have

$$\nabla f(x_0, y_0, z_0) = \lambda \, \nabla g(x_0, y_0, z_0) \tag{4}$$

for some constant λ. Hence to find candidates for the desired maximum, we should solve the following system equations, which come from considering the three coordinates of the gradient vectors, together with the original constraint equation:

$$\begin{aligned} \frac{\partial f}{\partial x}(x, y, z) &= \lambda \frac{\partial g}{\partial x}(x, y, z) \\ \frac{\partial f}{\partial y}(x, y, z) &= \lambda \frac{\partial g}{\partial y}(x, y, z) \\ \frac{\partial f}{\partial z}(x, y, z) &= \lambda \frac{\partial g}{\partial z}(x, y, z) \\ g(x, y, z) &= 0. \end{aligned} \tag{5}$$

The auxiliary unknown λ in this system is called the Lagrange multiplier for the problem.

Let us use the Lagrange multiplier method to solve the constrained maximization problem (3). The variables for this problem are Q_1, Q_2, and Q_3. The function we wish to maximize is $(KW_1 + KW_2 + KW_3)/(170 - 1.6 \cdot 10^{-6}\, Q_T^2)$, where we have divided by the constant factor for simplicity. The constraint is $Q_1 + Q_2 + Q_3 - Q_T = 0$. Note that in this maximization problem, Q_T is a constant (although we need to solve the problem for different values of Q_T). Exercise 1 asks you to check that equations (5) become

$$\begin{aligned} 0.1277 - 2(0.408 \cdot 10^{-4}\, Q_1) &= \lambda \\ 0.1358 - 2(0.469 \cdot 10^{-4}\, Q_2) &= \lambda \\ 0.1380 - 2(0.384 \cdot 10^{-4}\, Q_3) &= \lambda \\ Q_1 + Q_2 + Q_3 &= Q_T \end{aligned} \tag{6}$$

Exercise 2 asks you to check that this system has a unique solution, which is

$$\begin{aligned} Q_1 &= 0.3410\ Q_T - 75 \\ Q_2 &= 0.2967\ Q_T + 21 \\ Q_3 &= 0.3623\ Q_T + 54 \end{aligned} \tag{7}$$

Because of the upper and lower bounds on Q_1, Q_2, Q_3, this solution can only be attained for certain values of the total flow Q_T:

$$\begin{aligned} 250 \le Q_1 \le 1110 \quad &\text{implies} \quad 953 \le Q_T \le 3475 \\ 250 \le Q_2 \le 1110 \quad &\text{implies} \quad 772 \le Q_T \le 3670 \\ 250 \le Q_3 \le 1225 \quad &\text{implies} \quad 541 \le Q_T \le 3232 \end{aligned}$$

Hence our solution only works within the range $953 \le Q_T \le 3232$ cubic feet per second.

Exercises

1. Check that applying the Lagrange method to the three turbine problem does yield equations (6).
2. Solve the system of equations (6), and verify that the solution given in (7) is correct. What is λ?
3. For $Q_T = 2500$ cfs, how does the solution (7) tell us to divide the flow among the three turbines? How much power is produced? Check that some nearby distributions of the 2500 cfs flow produce less power, so that we do indeed have a maximum.

Other Configurations

The mathematical solution in the previous section is not a complete solution to the original problem. For one thing, what should be done when the total flow Q_T is outside the range [953, 3232]? For another, our solution assumes that we are running all three turbines. Perhaps even within the range [953, 3232], it might be better to run only one or two turbines.

If we were going to run just one or two turbines, we would want them to be the most efficient ones. The *efficiency* of a turbine is the number of kilowatts it produces per unit of flow, KW/Q. Since we only need to compare the turbines to each other, we can ignore the constant factor $C = (170 - 1.6 \cdot 10^{-6}\ Q_T^2)$. Figure 1 shows the graphs of the functions $KW/(Q \cdot C)$. As the flow increases, each turbine rises to a peak efficiency (see Exercise 4), and then declines slightly in efficiency if the flow continues to increase. We also see (Exercise 5) that

In [250, 430] the efficiency order is turbine 1, turbine 2, turbine 3.

In [430, 680] the efficiency order is turbine 1, turbine 3, turbine 2.

In [680, 1110] the efficiency order is turbine 3, turbine 1, turbine 2.

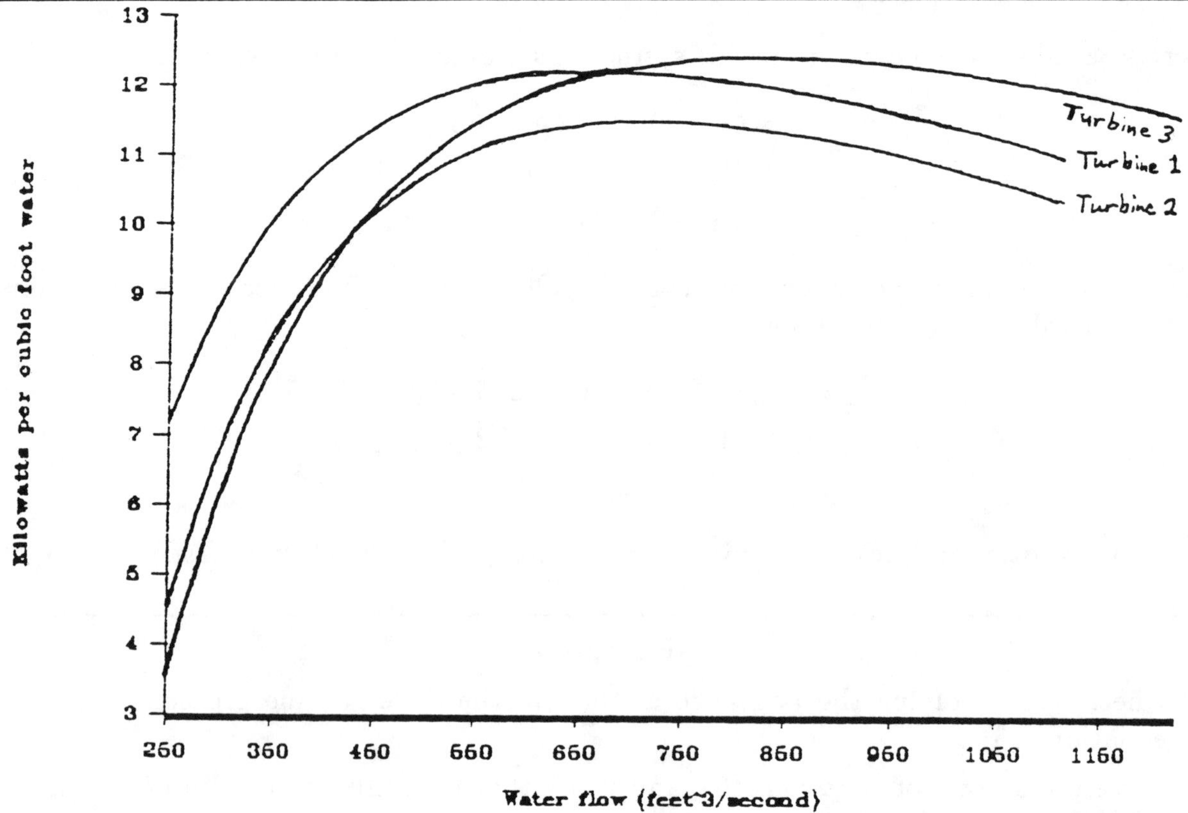

Figure 1. Efficiency of the Turbines.

If we run just one turbine, it should be turbine 1 in the flow range [250,680], and turbine 3 in the flow range [680,1225]. If we run two turbines, they should be turbines 1 and 2 or turbines 1 and 3, but we need to solve a constrained maximization problem in two variables to see how to divide the flow among the two turbines. It would be good practice (Exercise 6) for you to do this, using the fact that the Lagrange equations for the problem

$$\text{Maximize} \quad f(x,y) \quad \text{subject to} \quad g(x,y) = 0 \tag{8}$$

are

$$\begin{aligned} \frac{\partial f}{\partial x}(x,y) &= \lambda \frac{\partial g}{\partial x}(x,y) \\ \frac{\partial f}{\partial y}(x,y) &= \lambda \frac{\partial g}{\partial y}(x,y) \\ g(x,y) &= 0. \end{aligned} \tag{9}$$

The results are, for turbines 1 and 2:

$$\begin{aligned} Q_1 &= .5348\ Q_T - 46 \\ Q_2 &= .4652\ Q_T + 46 \\ 553 &\le Q_T \le 2162, \end{aligned} \tag{10}$$

for turbines 1 and 3:

$$\begin{aligned} Q_1 &= .4848\ Q_T - 65 \\ Q_3 &= .5152\ Q_T + 65 \\ 650 &\le Q_T \le 2252. \end{aligned}$$

The next step is to compare the power we can produce by using one, two or three turbines. For instance, suppose the total flow is $Q_T = 1000$. If we use one turbine, it should be turbine 3. If we use two, they should be turbines 1 and 2 or turbines 1 and 3. We compute the kilowatts produced in each configuration, using equations (2), (7) and (10):

Turbine 3 only:	$72.58 \cdot 168.4 = 12222$
Turbines 1 and 2:	$(33.78 + 32.66) \cdot 168.4 = 11188$
Turbines 1 and 3:	$(27.53 + 40.12) \cdot 168.4 = 11392$
Turbines 1, 2 and 3:	$(12.19 + 13.90 + 23.77) \cdot 168.4 = 8396$

We should use only turbine 3 for this value of Q_T. Is that a surprise? Looking at Figure 1 gives some insight. A flow of 1000 cfs lets turbine 3 operate near its peak efficiency. If we split the flow among two or three turbines, each has to operate significantly below peak efficiency.

There is one last situation we need to consider. What if the total flow is larger than 3232 cfs, which is the upper bound for our three turbine solution? Beyond that value, the solution would require that turbine 3 have flow above its maximum capacity of 1225 cfs. With a little thought, we see that we should use turbine 3 at full capacity, and then use the solution for turbines 1 and 2 to apportion the remaining flow $Q_T - 1225$. From equation (10) we see that this is good up to $Q_T = 2162 + 1225 = 3387$. Above that level, we should use turbine 3 and turbine 1 (the next most efficient turbine) at full capacity, and send the remaining flow to turbine 2. Of course, the plant cannot accept a flow greater than $1110 + 1110 + 1225 = 3445$ cfs.

Exercises

4. Find the flow Q which gives peak efficiency for each of turbines 1, 2 and 3.
5. Find the values of Q at which the graphs in Figure 1 cross.
6. Use the Lagrange method to solve the two turbine problem for turbines 1 and 2, and for turbines 1 and 3, checking that you get the answers in (10).
7. For each of the flows $Q_T = 600$ and $Q_T = 2200$, say which turbine configurations we need to check, and calculate the kilowatts which each of those configurations would produce. In each case, which configuration should we use?

Presenting the Solution

At this point, we need to give some thought to how we should present our solution. It must be in a form which can be used by the plant operator who controls flow to the turbines by valves and gates. In this context, a table giving the suggested distribution of flow for different values of Q_T would probably be more useful than mathematical formulas. Consultation with the plant operator suggests that, given the accuracy of the flow gauges, a table giving the distribution as Q_T increases in increments of 100 cfs would be satisfactory.

Table 1. Distribution of Flow for Optimal Power Production

Total Flow (cfs)	Flow to Turbine 1 (cfs)	Flow to Turbine 2 (cfs)	Flow to Turbine 3 (cfs)	Electrical Output (kilowatts)
250	250	0	0	1780
300	300	0	0	2670
400	400	0	0	4360
500	500	0	0	5900
600	600	0	0	7290
700	0	0	700	8590
800	0	0	800	9940
900	0	0	900	11140
1000	0	0	1000	12220
1100	0	0	1100	13160
1200	517	0	683	14350
1300	565	0	735	15710
1400	614	0	786	17000
1500	662	0	838	18210
1600	711	0	889	19340
1700	759	0	941	20400
1800	808	0	992	21380
1900	856	0	1044	22290
2000	606	615	779	23220
2100	641	644	815	24380
2200	675	674	851	25480
2300	709	703	888	26520
2400	743	733	924	27500
2500	777	763	960	28410
2600	812	792	996	29270
2700	846	822	1032	30060
2800	880	851	1069	30800
2900	914	881	1105	31470
3000	948	911	1141	32090
3100	982	941	1177	32650
3200	1016	970	1214	33140
3300	1064	1011	1225	33570
3400	1110	1065	1225	33920
3445	1110	1110	1225	34040

For values of Q_T starting at 250 and increasing in increments of 100 up to 3445, we need to compare the power produced by running one, two or three turbines according to our optimal flow distribution. It is possible, though messy, to do this analytically. However, it seems clear that this is a job for the computer. We can ask it to calculate the optimal distributions and compare the power outputs for the relevant turbine configurations (as we did in the previous section for $Q_T = 1000$ and you did in Exercise 7), and print out a table showing the distribution which produces the most power. The result is shown in Table 1. The electrical output has been rounded to the nearest 10 kilowatts. Notice that the final solution doesn't use all three turbines until the total flow reaches 2000 cfs. It was very important to realize that we should consider the possibility of using just one or two turbines.

Analyzing the Solution

Once we have presented our solution, it is natural to wonder how good it is. If we, as consultants, have been well paid for our work, the company may also be interested in knowing what our work is worth. Without our advice, one natural thing to do might have been to send equal flow to each turbine. Let us suppose that, without us, the company would have used the following strategy:

— in [250,500], use turbine 1,

— in [500,750], divide the flow equally between turbines 1 and 2,

— above 750, divide the flow equally among all three turbines.

For instance, when the total flow is 1000 cfs, each turbine would get 333 cfs. Using equations (2), we see that this would yield a power output of

$$(19.14 + 15.54 + 14.71)(168.4) = 7570 \text{ kilowatts.}$$

Our output of 12220 kilowatts is a 61% improvement—certainly worth our pay!

On the other hand, when the total flow is 2400 cfs, equal apportionment would send 800 cfs to each turbine, and the power output would be

$$(57.16 + 54.11 + 58.80)(160.78) = 27340 \text{ kilowatts.}$$

Our output of 27500 kilowatts is an improvement of less than 1%. Figure 2 compares the power output of our solution with that of the equal distribution strategy for all levels of Q_T.

Our solution is clearly superior in the range [500, 2000]. This is, of course, exactly the range where we use fewer turbines than we might, thereby allowing them to operate at higher efficiency levels. Above 2000, where we use all three turbines, the Lagrange solution is only slightly superior to the equal flow strategy. Figure 1 indicates the reason. For total flows above 2000 cfs, all turbines will be operating with flows above 600 cfs, and in this range the efficiency curves are both fairly level and fairly close together. There is only a little to be gained by switching flow from one turbine to another. However—we should emphasize the positive—under low flow conditions, our analysis offers a very significant improvement over a more naive strategy.

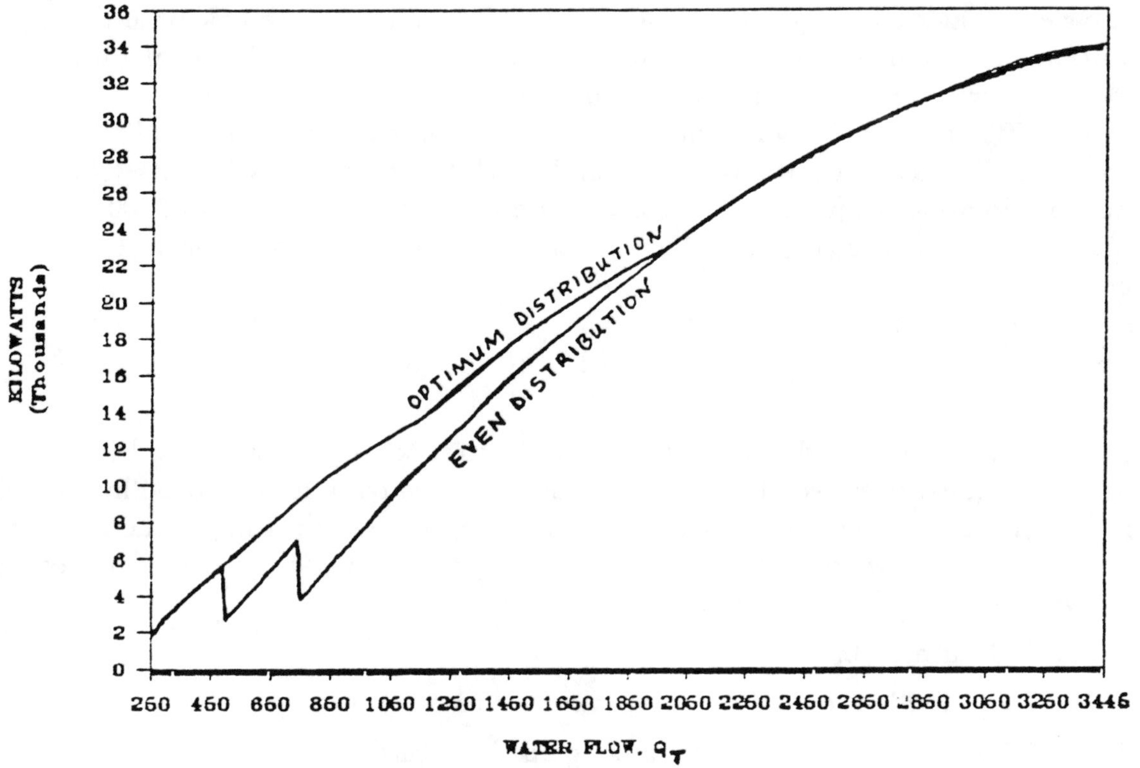

Figure 2. Optimal Distribution vs. Even Distribution.

Exercise

8. For the following flows, calculate the power which would be produced under the equal flow strategy, and compare it to the power produced under the optimal strategy. What is the percentage of improvement?

   a. $Q_T = 600$ b. $Q_T = 3000$.

A Discussion

The Lagrange multiplier technique used to solve this problem was developed by the Italian-French mathematician Joseph-Louis Lagrange in 1755, when Lagrange was 19 years old. Together with its generalizations to infinite dimensional problems in the "calculus of variations", due to Lagrange himself, and twentieth century generalizations to "optimal control theory", it is still the central technique for solving constrained optimization problems. The Tikhomirov reference is an extended tribute to its power.

However, the solution to Great Northern's problem involves much more than just applying a standard technique from calculus. This is typical of applied mathematics in industrial settings. First, we had to build a model—to formulate the problem mathematically. This required some background, in our case from the physics of fluid flow. It also required some experimental work to determine the loss of power due to friction in the supply tunnel (the term f in equation (1)) and, especially, the efficiency curves of the turbines. Those equations for KW_i were obtained by applying a statistical technique called "parabolic regression," which uses calculus to find a parabola which is the best fit to a set of data points.

The Lagrange multiplier technique then solved the three turbine problem. However, it was crucial to observe that we needed to check the possibility of using just one or two turbines, and indeed the benefits of final solution depended heavily on doing just that. Creativity and the careful use of common sense are both crucial to effective mathematical modeling.

Finally, notice the interplay between *analytic* techniques (Lagrange multipliers), *graphical* techniques (The efficiency curves in Figure 1 gave important insight into the problem), and *numerical* techniques (Table 1 presented our solution). This interplay pervades applied mathematics. If you are going to use mathematics to solve real problems, you should be willing to use all of them, and think about how they interrelate. For graphical and numerical work, computer assistance is often essential. However, this problem makes it abundantly clear that "putting it on the computer" can't substitute for modeling sense, creativity, and a powerful idea from calculus.

References

Agnew, Jeanne and Marvin Keener, eds., "Station hydro-turbine optimization," Department of Mathematics, Oklahoma State University.

Tikhomirov, V.M. (1991), *Stories about Maxima and Minima*, American Mathematical Society and Mathematical Association of America.

Figures 1 and 2 were generated by Matt Bilsbarrow, Danielle Carr and Tom Stimes, a team from my Mathematical Modeling class in 1987.

Answers to Exercises

2. $\lambda = .1338 - .2783 \cdot 10^{-4}\ Q_T$.
3. See Table 1 for the optimal flow distribution, which gives 28410 kilowatts, rounded to the nearest 10. A flow distribution of (750, 750, 1000) gives 28400 kw; a distribution of (800, 800, 900) gives 28380 kw.
4. For turbine 1, we need to maximize

$$\frac{-18.89}{Q} + .1277 - (.408 \cdot 10^{-4})\ Q.$$

Taking the derivative and setting it equal to zero gives

$$Q = \sqrt{\frac{18.89}{.408 \cdot 10^{-4}}} = 680.$$

For turbine 2, the maximum is at $Q = 723$; for turbine 3 it is at $Q = 839$.

5. The graphs for turbines 2 and 3 cross at $Q = 429$. The graphs for turbines 1 and 3 cross at $Q = 681$.

7.a. For turbine 1, Table 1 gives 7290 kw. The only other configuration we need to check is turbines 1 and 2, which would give $(13.14 + 14.67)(169.4) = 4710$ kw.

b. Table 1 gives 25480 for turbines 1, 2 and 3. The only other configuration we need to check is turbines 1 and 3, which would give $(68.10 + 83.19)(162.3) = 24550$ kw.

8.a. $(15.75 + 12.01)(169.4) = 4700$ kw. Our solution is 55% better.

b. $(68.01 + 64.39 + 72.58)(155.6) = 31895$ kw. Our solution is only 0.6% better.

PORTFOLIO THEORY

Author: Kevin J. Hastings, Knox College, Galesburg, IL 61401

Area of Application: finance

Calculus needed: differentiation, properties of the integral, Lagrange multipliers.

Related mathematics: statistical mean and variance.

The Portfolio Problem

'Buy low, sell high' may sound like wise advice, but it is certainly not easy to follow. If we open the daily paper to the stock quotations and see that currently a share of IBM stock sells for $103\frac{3}{4}$, down from $105\frac{1}{2}$ the day before, is the stock currently 'low' and ready for an advance, or is it 'high' and on the way down? Even when a stock is behaving in a relatively stable way over a time period, there is daily volatility which might make us nervous. We all want a decent return on our investments, but not at the cost of our feeling of security.

There are many kinds of information that might be used to predict the performance of stocks: general economic conditions, health of the industry which the stock represents, productivity of the company as reflected in its annual report, success of the company's competitors, and so on. It is not our purpose here to show how to make use of this type of information to play the stock market successfully. Rather we will suggest how, using stock quotations and a little bit of statistics, one can define and estimate the average rate of return on investment and the risk of investment. Then, with estimates in hand and an idea of the relative importance we give to riskiness as compared to return, we will use the technique of Lagrange multipliers to put together an optimal portfolio of investments.

The problem of choosing an optimal combination of investments is known in the economic literature as the *portfolio problem.* The 1990 Nobel Prize in Economics was awarded to Harry Markowitz, William Sharpe and Mertin Miller for developing the theory of portfolio optimization which is introduced in this module. For further reading, you might enjoy looking at Markowitz's original paper and Sharpe's book in the References.

Average Rate of Return

Suppose for concreteness that we are following four possible investments, whose market prices per share today are

$$P_1, P_2, P_3, P_4.$$

Let the share prices for the same stocks tomorrow be denoted by

$$Q_1, Q_2, Q_3, Q_4.$$

If we buy a share of stock 1 today, then the so-called *rate of return* on the investment, that is the gain in value per dollar paid, is

$$R_1 = \frac{Q_1 - P_1}{P_1}. \tag{1}$$

Rates of return on the other stocks are defined analogously. It is obvious that we should prefer high rates of return to low rates of return. Now the rate of return on stock 1 is unknown to us today; we must wait until tomorrow if we want to know it. But then we have missed our investment opportunity. The way of getting around this dilemma is to suppose that, at least over a short time period, there is a constant but unknown 'average' rate of return on stock 1, call it ρ_1, about which R_1 fluctuates as days pass. A reasonable way to estimate this theoretical rate of return is to follow the stock over as many days as possible, compute from the price data the succession of rates of return as in formula (1), and average those.

Date	WalMart	Rate of Return	Goodyear	Rate of Return	Honda	Rate of Return	ComEd	Rate of Return
June 5	42		30.125		20.25		36.5	
June 6	41.375	-1.5%	31.375	4.1%	20.375	0.6%	36.75	0.7%
June 7	41.875	1.2%	32.125	2.4%	20.5	0.6%	36.5	-0.7%
June 8	41.625	-0.6%	33.25	3.5%	20.375	-0.6%	36.375	-0.3%
June 11	42.125	1.2%	33.25	0.0%	20	-1.8%	36.625	0.7%
June 12	43.125	2.4%	32.125	-3.4%	20.625	3.1%	37.375	2.0%
June 13	42.5	-1.4%	33.25	3.5%	20.625	0.0%	37.25	-0.3%
June 14	42.625	0.3%	33.125	-0.4%	21.125	2.4%	37.5	0.7%
June 15	43.375	1.8%	33.5	1.1%	21.375	1.2%	37.875	1.0%
June 18	42.75	-1.4%	34.875	4.1%	21.125	-1.2%	37.625	-0.7%
June 19	42.5	-0.6%	34	-2.5%	20.875	-1.2%	37.375	-0.7%
June 20	42.375	-0.3%	34.125	0.4%	20.625	-1.2%	37.125	-0.7%
June 21	43	1.5%	34.25	0.4%	21.25	3.0%	37.375	0.7%
June 22	43.375	0.9%	34.25	0.0%	21.5	1.2%	37.625	0.7%
June 25	42.375	-2.3%	33.875	-1.1%	21	-2.3%	36.375	-3.3%
June 26	42.25	-0.3%	33.5	-1.1%	21.375	1.8%	36.75	1.0%
June 27	43.375	2.7%	33.125	-1.1%	21.375	0.0%	37.125	1.0%
Mean		.212%		.620%		.352%		.113%
Var		.022%		.054%		.029%		.015%

Table 1.

For example, consider Table 1. We have recorded the prices of the common stocks of four companies trading on the New York Stock Exchange: WalMart, Goodyear, Honda, and Commonwealth Edison. We followed the stocks over a period of seventeen trading days. The numbers in the rate of return columns are obtained by dividing the difference (current price minus previous price) by the previous price. Taking the simple arithmetical average of those observed rates of return gives the estimates of $\rho_1, \rho_2, \rho_3,$ and ρ_4 listed in the row labeled 'Mean'. We will discuss the row labeled 'Var' later. In Figure 1 are graphs of the rates of return of two of the stocks as functions of trading day.

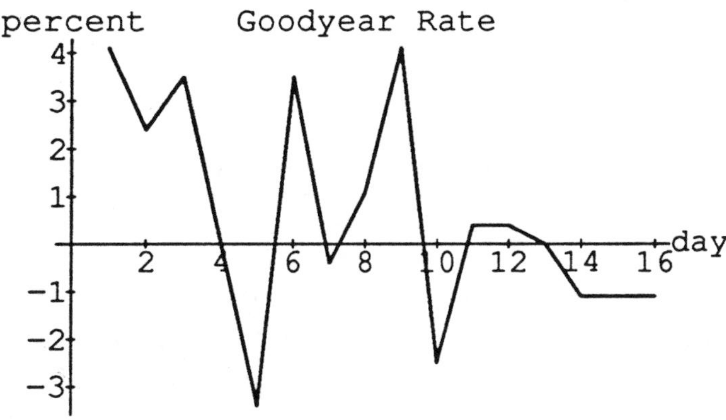

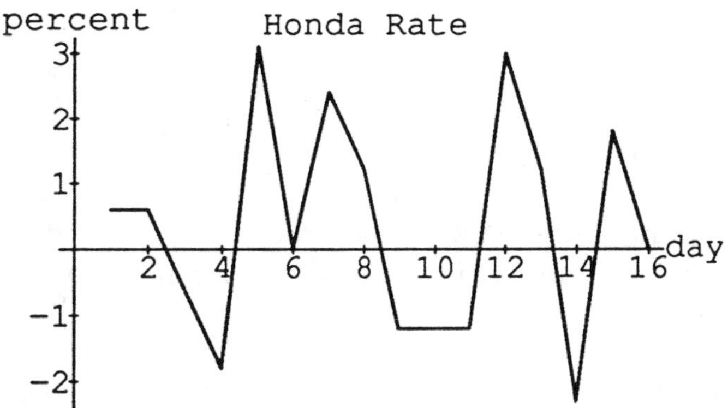

Figure 1.

As the table and the graphs suggest, there is variability in the rates of return, sometimes a lot of it. Typically, risky investments which have high average rates of return also have high risks. If this was not the case, then investors would all flock to high-reward, low-risk investments, and market imbalances would occur. To increase safety, investors tend to diversify, that is, to spread their wealth among several opportunities, hence constructing a collection or *portfolio* of investments. If the investor has total wealth W, then the decision to be made is what fraction w_i of that wealth is to be devoted to investment i, for i=1,2,3,4. The total wealth invested in investment i is $w_i \cdot W$, and since the expected rate of return per dollar invested is ρ_i, the expected dollar return from our investment in investment i is

$$\rho_i \cdot w_i \cdot W \quad . \tag{2}$$

Summing over all four investments, the total dollar return on the portfolio is expected to be

$$\rho_1 \cdot w_1 \cdot W + \rho_2 \cdot w_2 \cdot W + \rho_3 \cdot w_3 \cdot W + \rho_4 \cdot w_4 \cdot W \quad . \tag{3}$$

Since we invested W dollars, the rate of return for the portfolio per dollar invested is

$$\rho = \frac{\rho_1 \cdot w_1 \cdot W + \rho_2 \cdot w_2 \cdot W + \rho_3 \cdot w_3 \cdot W + \rho_4 \cdot w_4 \cdot W}{W} = \rho_1 \cdot w_1 + \rho_2 \cdot w_2 + \rho_3 \cdot w_3 + \rho_4 \cdot w_4 \, . \tag{4}$$

If we did not care about riskiness, then to maximize the rate of return (4) on the portfolio, we would find the largest of the individual rates of return, say ρ_2, and devote $w_2 = 1 = 100\%$ of our wealth to that investment, leaving $w_i = 0 = 0\%$ of our wealth to the other investments, $i = 1, 3, 4$. The portfolio problem becomes much more interesting if we take riskiness into account. In the next section we take a look at the idea of risk.

Exercises

1. Choose two of your favorite companies and follow the progress of their prices on the New York Stock Exchange for a few days. Obtain estimates of their average rates of return. If you had $10,000 to invest in these stocks, what would you do, and why? Does it seem to you that a few days worth of information is sufficient to make a decision? Why or why not?

2. Suppose that there are three stocks, with rates of return $\rho_1 = 5\%, \rho_2 = 10\%, \rho_3 = 15\%$. Suppose also that because the third stock is very speculative, you want no more than half as much wealth in it as in stock 1. What is the maximum portfolio rate of return that you can achieve? (Hint: Express the problem in terms of portfolio weights w_1, w_2 only, and look carefully at the region of points in the $w_1 - w_2$ plane that are legal under the problem conditions. Where in that region is the portfolio rate of return largest?)

Risk and Risk Aversion

We mentioned in the last section that a rate of return R on a stock is not something that can be perfectly predicted; hence there is riskiness in investment. We also assumed that there is an unknown theoretical average (or *mean*) rate of return ρ which can be estimated from a sample of data by the *sample mean*, which will now be denoted by $\overline{R}$. In general, if $a_1, a_2, \ldots, a_n$ are measurements of some quantity (such as the rate of return for a given stock) the sample mean of the measurements is

$$\overline{a} = \frac{a_1 + a_2 + \ldots + a_n}{n} = \frac{1}{n}\sum_{i=1}^{n} a_i. \tag{5}$$

Let us now introduce a notion of theoretical risk that is called *variance* by statisticians, which is also easily estimable from a sample of data.

The *variance* $Var(R) = \sigma^2$ (read 'sigma squared') of an uncertain rate of return is the average squared difference between the rate of return R and its mean ρ. The squaring ensures that both positive and negative differences contribute positively to the variance. You can find out more about mean and variance in the Hogg and Tanis reference, as well as many other sources. It is enough for us to know intuitively that the variance will be large when values of R far from the mean ρ occur with high likelihood, and the variance will be small when R is very likely to be close to ρ. In this way, variance becomes a reasonable measure of risk.

A simple way of estimating the theoretical variance σ^2 is to take a sample, compute the sample mean for that sample, and then average the squared differences of the sample values from the sample mean. If there are n sample values, it would seem that in averaging we should divide the total by n, but technical reasons indicate that this method underestimates the variance slightly. It is more common to divide by $n-1$ instead. In general, if $a_1, a_2, \ldots, a_n$ are measurements of some quantity (such as the rate of return for a given stock) the *sample variance* S^2 of the measurements is

$$S^2 = \frac{(a_1 - \overline{a})^2 + (a_2 - \overline{a})^2 + \ldots + (a_n - \overline{a})^2}{n-1} = \frac{1}{n-1}\sum_{i=1}^{n}(a_i - \overline{a})^2. \tag{6}$$

When the data is widely spread out from the sample mean, the sample variance will be large.

For the rate of return data in the table in the last section, the sample variance of rate of return for the sixteen observations of the Walmart stock is:

$$\frac{1}{15}[(-1.5 - .212)^2 + (1.2 - .212)^2 + (-0.6 - .212)^2 + \ldots + (2.7 - .212)^2] = .022$$

The sample variance was calculated for each of the stocks, and the results are displayed in the row labeled "Var" at the bottom of Table 1.

In the portfolio problem we will need to consider the risk or variance of the portfolio's actual rate of return, which, by analogy with formula (4), is

$$R_p = w_1R_1 + w_2R_2 + w_3R_3 + w_4R_4. \tag{7}$$

What is the variance of this random quantity? First, it happens to be the case that when a random quantity R is multiplied by a constant w, its theoretical mean is also multiplied by w, and its theoretical variance is multiplied by w^2. The sample mean and variance also have these properties, as you will show in Exercise 3. Second, though we are not prepared to show it here, when several random quantities satisfy a property called ***stochastic independence***, meaning roughly that the value of one does not affect the probabilistic laws of the others, then the variance of the sum is the sum of the variances of the individual terms in the sum. These results allow us to write

$$\begin{aligned} Var(R_p) &= Var(w_1R_1 + w_2R_2 + w_3R_3 + w_4R_4) \\ &= Var(w_1R_1) + Var(w_2R_2) + Var(w_3R_3) + Var(w_4R_4) \\ &= w_1^2Var(R_1) + w_2^2Var(R_2) + w_3^2Var(R_3) + w_4^2Var(R_4) \\ &= w_1^2\sigma_1^2 + w_2^2\sigma_2^2 + w_3^2\sigma_3^2 + w_4^2\sigma_4^2. \end{aligned} \tag{8}$$

Thus, by estimating the individual rate of return variances σ_i^2 by sample variances S_i^2, we have an explicit quadratic expression for the portfolio variance as a function of the portfolio weights.

A brief digression is in order here. It is not too surprising that risks should be additive, though it requires an assumption. Interestingly, when the independence assumption is not satisfied, extra terms appear on the right of (8) which can be negative if one stock tends to go down as another goes up. These negative terms, arising from correlations among stocks, actually reduce portfolio risk as compared to the independent case. Intuitively, by combining two stocks which tend to move in opposite directions one is protected against losses in one by gains in the other. This is the heart of diversification in investment problems.

At this point we have provided a notion of average rate of return on a portfolio $\rho = \sum w_i\rho_i$ (equation(4)) and variance of the portfolio $\sigma^2 = \sum w_i\sigma_i^2$. We would like to choose investment fractions w_i which maximize the average rate of return and minimize the risk. In general, these two goals are incompatible, since the most profitable investments tend to be the riskiest (see Table 1). Hence we need to balance these two goals somehow.

One way to do this is to maximize a kind of net value of the portfolio, $\rho - a\sigma^2$, where a is a ***risk aversion factor*** which measures your reluctance to take risks. For example $a = 0$ means that you don't mind taking risks at all; your only goal is to maximize average rate of return. If $a = 10$, then an increase of one unit in risk would have to be offset by an increase in 10 units of rate of return in order for you to value the investment in the same way. A larger value of a means more worry about risk.

If an investor with risk aversion a is presented with two portfolios having average rates of return ρ and $\hat{\rho}$ and variances σ^2 and $\hat{\sigma}^2$, then the investor will be indifferent between the two portfolios if

$$\rho - a\sigma^2 = \hat{\rho} - a\hat{\sigma}^2 \ . \tag{9}$$

Solving for a, we see that

$$a = \frac{\rho - \hat{\rho}}{\sigma^2 - \hat{\sigma}^2}. \tag{10}$$

In assuming that an investor has risk aversion a, we are assuming that this quotient is the same for all pairs of investments between which the investor is indifferent. Equation (10) then gives us a way to estimate an investor's risk aversion. We will assume that a is constant and known, and proceed to solve the portfolio problem.

Exercises

3. a. Show that if each observation in a sample $a_1, a_2, \ldots, a_n$ is multiplied by a constant w, then the sample mean is also multiplied by w.

b. Show that if each observation in a sample $a_1, a_2, \ldots, a_n$ is multiplied by a constant w, then the sample variance is multiplied by w^2.

4. For the stocks you followed in Exercise 1, estimate the variances of their rates of return. Does this information affect your response to the question in Exercise 1?

5. What is your risk aversion constant if you are indifferent between two portfolios, one of which has mean return 5% and variance 1%, and the other has mean return 8% and variance 4%?

6. When stock returns are dependent on one another, the variance of a sum is not just the sum of the variances, but also includes a *covariance* term, e.g. for two stocks:

$$Var(w_1R_1 + w_2R_2) = w_1^2Var(R_1) + w_2^2Var(R_2) + 2w_1w_2Cov(R_1, R_2)$$

The covariance measures the degree of dependence between the two returns. Suppose that $Var(R_1) = 2, Var(R_2) = 1$ and $Cov(R_1, R_2) = -1$. By how much is the portfolio variance of an equally balanced portfolio reduced, as compared to the independent case? What values of w_1 and w_2 minimize the variance of the portfolio? Are they different weights than those which minimize variance in the independent case?

Solving the Optimization Problem

Agreeing to maximize $\rho - a\sigma^2$ reduces the portfolio problem to the constrained optimization problem

$$\begin{aligned} \text{Maximize} \quad f(w_1, w_2, w_3, w_4) =& \rho_1w_1 + \rho_2w_2 + \rho_3w_3 + \rho_4w_4 \\ & - a(w_1^2\sigma_1^2 + w_2^2\sigma_2^2 + w_3^2\sigma_3^2 + w_4^2\sigma_4^2) \\ \text{subject to} \quad g(w_1, w_2, w_3, w_4) =& w_1 + w_2 + w_3 + w_4 - 1 = 0. \end{aligned} \tag{11}$$

The Lagrange multiplier technique tells us that we may find possible maximum points by solving simultaneously the equations

$$\nabla f = \lambda \nabla g, \qquad g = 0, \tag{12}$$

where λ is a new variable and ∇ indicates the gradient operator. The partial derivatives of f and g are easy to find:

$$\frac{\partial f}{\partial w_i} = \rho_i - 2a\sigma_i^2 w_i, \quad \frac{\partial g}{\partial w_i} = 1, \qquad i = 1, 2, 3, 4. \tag{13}$$

These computations result in a system of equations in the unknowns λ, w_1, w_2, w_3 and w_4:

$$\begin{aligned} \rho_1 - 2a\sigma_1^2 w_1 &= \lambda \\ \rho_2 - 2a\sigma_2^2 w_2 &= \lambda \\ \rho_3 - 2a\sigma_3^2 w_3 &= \lambda \\ \rho_4 - 2a\sigma_4^2 w_4 &= \lambda \\ w_1 + w_2 + w_3 + w_4 &= 1. \end{aligned} \tag{14}$$

By equating λ in equations 2, 3, and 4 with λ in the first equation, we can find w_2, w_3 and w_4 in terms of w_1:

$$\begin{aligned} \rho_1 - 2a\sigma_1^2 w_1 &= \rho_2 - 2a\sigma_2^2 w_2 \\ \text{gives} \qquad w_2 &= \frac{(\rho_2 - \rho_1) + 2a\sigma_1^2 w_1}{2a\sigma_2^2} \end{aligned} \tag{15}$$

and similarly

$$w_3 = \frac{(\rho_3 - \rho_1) + 2a\sigma_1^2 w_1}{2a\sigma_3^2} \tag{16}$$

$$w_4 = \frac{(\rho_4 - \rho_1) + 2a\sigma_1^2 w_1}{2a\sigma_4^2}. \tag{17}$$

The last condition of system (14) lets us solve for w_1:

$$\begin{aligned} 1 &= w_1 + w_2 + w_3 + w_4 \\ &= w_1 + \frac{\sigma_1^2}{\sigma_2^2} w_1 + \frac{(\rho_2 - \rho_1)}{2a\sigma_2^2} + \frac{\sigma_1^2}{\sigma_3^2} w_1 + \frac{(\rho_3 - \rho_1)}{2a\sigma_3^2} + \frac{\sigma_1^2}{\sigma_4^2} w_1 + \frac{(\rho_4 - \rho_1)}{2a\sigma_4^2}, \\ \text{so that } w_1 &= \frac{1 - \frac{(\rho_2 - \rho_1)}{2a\sigma_2^2} - \frac{(\rho_3 - \rho_1)}{2a\sigma_3^2} - \frac{(\rho_4 - \rho_1)}{2a\sigma_4^2}}{1 + \frac{\sigma_1^2}{\sigma_2^2} + \frac{\sigma_1^2}{\sigma_3^2} + \frac{\sigma_1^2}{\sigma_4^2}}. \end{aligned} \tag{18}$$

Formula (18) is now an explicit formula for w_1 in terms of the parameters of the problem. The other weights w_2, w_3, w_4 can be computed from w_1 by (15), (16), and (17). We have done the numerical computations for several values of the risk aversion a and the estimated means and variances of return from Table 1, and we find the following optimal values for the portfolio weights:

Risk Aversion:	2	4	6	8	10	12
ComEd	-1.83	-0.71	-0.34	-0.15	-0.04	0.03
WalMart	-0.12	0.08	0.14	0.18	0.19	0.21
Honda	1.11	0.66	0.51	0.44	0.39	0.35
Goodyear	1.84	0.98	0.69	0.54	0.46	0.40

Table 2.

There are a few interesting things to be noted about the weights in Table 2. First, there are some negative values, particularly for the low-return, low-risk stock Commonwealth Edison. Nothing in the mathematics that we did prevented this from happening, and until we actually performed the analysis, we could not have guessed that this would happen. Is our model flawed? Not necessarily, because it is actually possible to hold negative wealth in an asset, by taking what is called in finance circles a 'short position' in that asset. An investor can do this by contracting to acquire the proceeds of sale of the asset as if the investor actually owned it, then paying back at a later time by 'buying' the asset back at current market price, thus reversing the order in the usual buy-sell cycle. The investor profits if the stock price goes down, so that the 'purchase' price is less than the 'selling' price. If this opportunity is not available or not appealing to us, the original model would have to be redesigned as a constrained optimization problem with non-negativity constraints on the variables w_i. This would take us farther afield into the subject of mathematical programming than we would like to travel, but it is an important extension to the problem.

We will see in the next section that if in addition to risky investments like stocks ($\sigma^2 > 0$), we consider a *risk-free* investment opportunity like a bond or savings account ($\sigma^2 = 0$), then, under reasonable conditions, the weights for the risky investments will always be positive. In other words, in this more realistic context, the problem of negative weights for stocks does not arise.

Another interesting aspect of the data in Table 2 is much more foreseeable. We have ordered the stocks top to bottom from least risky to most risky. For these stocks, the mean returns also increase from top to bottom. As risk aversion increases, more of the wealth shifts from the more speculative stocks (at least in terms of our limited data), Honda and Goodyear, to the safer stocks, ComEd and WalMart.

For another unexpected result in the case where one of the available stocks seems to be strictly preferable to another, see Exercise 8.

Exercises

7. Suppose that there are three mutually independent stocks, with $\rho_1 = 5\%, \sigma_1^2 = 0\%, \rho_2 = 8\%, \sigma_2^2 = 2\%, \rho_3 = 12\%, \sigma_3^2 = 4\%$. What is the optimal portfolio for an investor whose risk aversion is $a = 2$? Risk aversion $a = 3$?

8. a. In the process of obtaining the data on the four stocks used as examples, I traced the activity of several others and noticed a surprising result. Over the same seventeen day trading period, sixteen observations of rates of return on Exxon, Hormel, and Abbott Laboratories gave mean rates of return .085, .134, and .179 respectively, and variances .011, .029, .025 respectively. Then Abbott Labs appears to be strictly better than Hormel, in the sense of having a higher expected return, with lower risk. One would guess that an optimal portfolio consisting of these stocks, together with, say Goodyear as a fourth stock, would not include Hormel at all. Check that in actuality this is not the case using a risk aversion of $a = 4$.

b. If you are curious and ambitious, write general expressions for the value function $\rho - a\sigma^2$ for a portfolio that uses all stocks, and for a second portfolio like the first, except that the weight formerly given to stock 2 (Hormel) is lumped together with the weight for stock 3 (Abbott Labs). Try to see what it is about the two values that makes the first better than the second, and think about what your study implies about diversification of investments.

The Portfolio Separation Theorem

Before closing, let us use our results to prove a beautiful and important theorem of investment economics.

Portfolio Separation Theorem. *Suppose one possible investment in a portfolio is risk-free (for example a savings account or a bond). Then the ratios of the optimal weights of the other investments is independent of the investor's risk aversion.*

Proof: Suppose that investment 1 is risk-free, so that $\sigma_1^2 = 0$. Then formulas (15), (16), and (17) simplify to

$$w_2 = \frac{\rho_2 - \rho_1}{2a\sigma_2^2}, \qquad w_3 = \frac{\rho_3 - \rho_1}{2a\sigma_3^2}, \qquad w_4 = \frac{\rho_4 - \rho_1}{2a\sigma_4^2}. \tag{19}$$

The ratios of these weights are indeed independent of the risk aversion a. For example,

$$\frac{w_3}{w_2} = \frac{(\rho_3 - \rho_1)/2a\sigma_3^2}{(\rho_2 - \rho_1)/2a\sigma_2^2} = \frac{(\rho_3 - \rho_1)/\sigma_3^2}{(\rho_2 - \rho_1)/\sigma_2^2}$$

and the theorem is proved.

Another way of stating the result of the portfolio separation theorem is that the portion of the total wealth in risky investments which is devoted to each single risky investment does not depend on the risk aversion, e.g.

$$\frac{w_2 W}{w_2 W + w_3 W + w_4 W} = \frac{\frac{\rho_2 - \rho_1}{\sigma_2^2}}{\frac{\rho_2 - \rho_1}{\sigma_2^2} + \frac{\rho_3 - \rho_1}{\sigma_3^2} + \frac{\rho_4 - \rho_1}{\sigma_4^2}}.$$

This means that any investor working with the same knowledge about rates of return and variance, and using the type of objective that we are using, will hold these investments in the same proportion to each other, regardless of the degree of risk aversion. These constant relative proportions describe what economists call the ***market portfolio***. The role of the risk aversion a is only to determine what proportion of wealth is devoted to the risk-free investment, not the relative mixture of the risky investments.

Finally, notice that if all of the risky investments have higher rates of return than the risk-free investment, i.e. $\rho_2, \rho_3, \rho_4 > \rho_1$, then by (19) the weights of the risky investments w_2, w_3 and w_4 are all positive. This is the result referred to in the previous section's discussion of negative weights. The weight w_1 of the risk-free investment may still be negative (see Exercise 9) if our risk aversion is small. A negative w_1 in this case has the interpretation that we should borrow money to buy more risky investments.

Exercises

9. Suppose your investment opportunities are stocks in Exxon, Hormel, and Abbott Labs (see the data in Exercise 8), and a bond with $\rho = .050$ and $\sigma^2 = 0$. What is your optimal portfolio if your risk aversion is $a = 4$? $a = 8$? $a = 12$? (What do you suppose it means if the weight you should invest in the bond is negative?) What is the market portfolio of the three stocks?

10. Consider a portfolio problem with five investments, the first risk-free, and with the extra constraint that investments 2 and 3 must have equal portions of the total wealth. Does a portfolio separation theorem still hold in this case? (Note: For this problem you will need an extra Lagrange multiplier for the extra constraint.)

References

Hogg, Robert V., and E.A. Tanis (1988), *Probability and Statistical Inference, 3rd ed.*, Macmillan.

Markowitz, Harry (1952), "Portfolio selection," *Journal of Finance* 7: 77-91.

Sharpe, William. (1970), *Portfolio Theory and Capital Markets*, McGraw-Hill.

Samuelson, Paul A. (1983), *Foundations of Economic Analysis*, Harvard.

Answers to Exercises

2. Invest all in stock 2; 10%.

5. The risk aversion is 1.

6. The variance in the dependent case is only a third of what it is in the independent case. The minimum variance in the dependent case occurs for $w_1 = 2/5, w_2 = 3/5$, and in the independent case it occurs for $w_1 = 1/3, w_2 = 2/3$.

7. For $a = 2$, $w_1 = 3/16, w_2 = 3/8, w_3 = 7/16$. For $a = 3$, $w_1 = 11/24, w_2 = 1/4, w_3 = 7/24$.

8. The optimal weights come out to be $w_1 = -.455, w_2 = .0387, w_3 = .2699, w_4 = 1.146$. For the second question, the issue of which is better turns on the size of $(w_2 + w_3)^2$ relative to $w_2^2 + w_3^2$.

9.

Risk aversion	4	8	12	Market portfolio
Bond	−.41	.30	.53	
Exxon	.40	.20	.13	.28
Hormel	.36	.18	.12	.26
Abbott	.65	.32	.22	.46

Investing a negative amount in the bond might be interpreted as borrowing to buy more stock.

10. A separation theorem does hold. Formulas analogous to (16) still apply to assets 4 and 5, and the weights $w_2 = w_3$ also satisfy an equation like (16), with mean ρ_3 replaced by the average of the two asset means, and the variance σ_3^2 replaced by the average of the two asset variances.